Other books by John Horner

Digging Dinosaurs, with James Gorman
Maia: A Dinosaur Grows Up, with James Gorman
Digging Up Tyrannosaurus rex, with Don Lessem

Other books by Don Lessem

Dinosaurs Rediscovered (formerly *Kings of Creation*)
Life Is No Yuk for the Yak
The Worst of Everything

THE COMPLETE
T. rex

John R. Horner and Don Lessem

A TOUCHSTONE BOOK
PUBLISHED BY

SIMON & SCHUSTER

NEW YORK   LONDON   TORONTO   SYDNEY   TOKYO   SINGAPORE

TOUCHSTONE
Rockefeller Center
1230 Avenue of the Americas
New York, New York 10020

First Touchstone Edition 1994

TOUCHSTONE and colophon are
registered trademarks of Simon & Schuster Inc.

Designed by Deborah Perugi
Manufactured in the United States of America

10 9 8 7 6 5 4 3 2 1

Library of Congress Cataloging-in-Publication Data

Horner, John R.
The complete *T. rex:* how stunning new discoveries are
changing our understanding of the world's most famous
dinosaur / John R. Horner and Don Lessem.
        p.   cm.
Includes bibliographical references and index.
1. *Tyrannosaurus rex.*   I. Lessem, Don.   II. Title.
QE862.S3H66   1993
567.9'7–dc20   93–211   CIP
ISBN: 0-671-74185-3
        0-671-89164-2 (Pbk.)

ACKNOWLEDGMENTS

This book was made possible by the work of a host of others who have researched *Tyrannosaurus rex* and its world and informed our studies at the Museum of the Rockies. Bill Abler, Ken Carpenter, Bill Clemens, Phil Currie, Jim Farlow, Don Glut, Leo Hickey, Kirk Johnson, Ralph Molnar, Mark Norell, and Scott Wing kindly reviewed the manuscript in part or whole in its roughest form, and Bob Bakker, Harley Garbani, and Peter Larson provided valuable comments along the way.

Equally deserving of thanks are the members of the Museum of the Rockies staff, Pat Leiggi, Carrie Ancell, Bob Harmon, and the rest of the field crew, preparators, and graduate students—a uniquely skilled group.

Special thanks go also to designers Deb Perugi and Adam Kelley and the many talented artists who have supplied work for this book, including Kris Ellingsen, Brian Franczak, Doug Henderson, Kit Mathers, Pat Ortega, Greg Paul, and Matt Smith; Photographer Bruce Selyam; filmmaker Mark Davis; NOVA producer Paula Apsell; publishing guru Bernice Colt; copyeditor Mary Anne Stewart, indexer Miriam Witlin, proofreader Sydney Fishman, and know-it-all George Olshevsky.

Thanks, too, to editor Bob Bender and his able assistant, Johanna Li. And last, but not least, we thank our families for their patience with us when we're messing around with dinosaurs instead of them. We're glad we can do both.

to the Wankel family

CONTENTS

T. REX BATTLING
AN *EDMONTONIA*
BY BRIAN FRANCZAK

THE COMPLETE *T. REX*

WILL THE REAL *TYRANNOSAURUS REX* PLEASE STAND UP?

*W*ESTERN NORTH AMERICA, 65 MILLION YEARS AGO. The last of the dinosaurs are on the move. They're roaming inland and along the margins of a big, shallow inland sea. These animals are huge, many more than twenty feet long. But there are far fewer kinds of them than there were 10 million years before, when the seas were wider and the temperatures milder.

Huge herds of horned dinosaurs and giant duckbills tromp across the land, heading north in summer, south in winter. They munch on ferns and flowering plants. As they go, they're bleating and honking to each other. The noise can be deafening and the smell overpowering. Here and there smaller dome-headed dinosaurs browse, males butting heads in loud collisions. Squat armored dinosaurs, the size of small tanks, lumber about. They're scarfing up low-growing plants. Scurrying about beneath the dinosaurs' feet are our ancestors, insect-munching mammals. The mammals are snacks for nimble, man-sized dinosaurs. Other, bigger dinosaur predators are on the loose, among them a pygmy tyrannosaur. This killer is only fifteen feet long, with quick feet and good eyesight.

But all these animals live in terror of one of the greatest carnivores ever—*Tyrannosaurus rex.* Suddenly it approaches, rushing in from hiding in the underbrush, carrying its tail high as its thickly muscled legs pump in long, narrow, swift strides. The herds scatter, exposing the young, the old, and the frail, which lag behind. The hunter corners one sickly *Triceratops*, which turns, bucks its head, and flashes its menacing horns at the predator. The killer's huge maw opens, revealing gleaming serrated fangs the size and shape of bananas, poised to tear into a hapless adversary.

Lowering its head, *T. rex* chomps down on the back of the horned dinosaur. *T. rex*'s stubby, powerful claws lock into the tough hide of the victim, securing it while jaws and teeth shake the flailing prey, tearing away huge, bloody hunks of flesh.

This *T. rex* is fast, and nimble. It might hunt in packs as well as alone, butting heads with its rivals. To kill, it leaps out to kick or pinion its prey with a massive hind leg or crunch it with a lethal bite. It is one fleet-footed, mean, killing son-of-a-gun.

Another day, another imagined view of *T. rex*. This is an equally graceful monster. It, too, is fast enough to catch prey, but rather than get into a nasty struggle to kill its dinner, it uses its monstrous jaws to tear huge hunks of meat from the many carcasses littering the landscape. It's the vulture of its day: a huge, efficient leftover-eater.

Which of these *T. rex*es comes closer to the truth? Either one, depending on how you interpret the evidence. Each vision of *T. rex* is based on reasonable inferences, unlike the antiquated vision of *T. rex* as a fat and sluggish cold-blooded reptile. Each of these modern scenarios is based on the best information we have about *T. rex*. That information is growing fast. In the past few years, we've found out more about *T. rex* then we ever knew before. My co-workers and I at the Museum of the Rockies dug up one of the two most complete *T. rex*es of all in 1990 and found some surprises, among them that *T. rex*'s front arms, long imagined to be puny and useless, weren't weak after all.

We'll tell you here what we paleontologists now know about *T. rex*, how we dug out bones and figured out how *T. rex* looked and acted. But for all we know about *T. rex*, much more remains a mystery. Some questions may never be answered. Others require us to speculate, to make reasonable guesses based on the good information we do have.

Some people don't like to hear scientists use the word *speculate*. They think science is all hard data and certain answers. It's not, especially when you've got a science with as many gaps and as little data as we have for the evolution and behavior of dinosaurs.

It's as if we're detectives investigating a murder, only we weren't there, and we don't have the culprit or much of the evidence. Speculation is healthy, very healthy, as long as it is grounded in evidence. Where you get into trouble is when you start believing the speculation simply because it appeals to you.

Dinosaurs are awesome. All of us would really like to know what they looked like, how they acted and moved. That leads to speculation. That doesn't mean any speculation we make is good science. If we do speculate, we need to identify it as speculation, not fact. And in our speculation we need to try for the neatest, most parsimonious solution to fit the facts we do have.

Since there is speculation in all science, it is important to understand that just because one particular scientist says this is what happened, it isn't necessarily so. That's only his or her speculation.

Science is constantly changing. We're always learning new things. The first person who comes up with an idea always has the least information that will ever be known about that idea. But we can never solve all the mystery.

We'll never know the absolute, complete truth about dinosaurs. Certainly not about *T. rex*. In the first ninety years that we knew of *T. rex*'s existence, scientists had uncovered only eight skeletons of the animal, none of them more than 60 percent complete.

Now that's all changed. Here's how and why.

CHAPTER 1

HOW TO FIND A *T. REX*

ASTERN MONTANA, June 1990. I'm on my knees on the top of a bare, dusty hill. The hill is on the muddy shore of a reservoir in the middle of mile after mile of badlands. Ten expert dinosaur diggers are working next to me. With air hammers, picks, even toothbrushes, we're picking away at a skeleton bigger than all of us. We're breaking up and carting away tons of sandstone to get at the brown bones of the biggest killer anyone's ever found. It's late afternoon, and still it's 110 degrees in the shade. Only there isn't any shade. No wonder this place is called Hell Creek.

None of us is stopping work. We've got only a month to uncover this dinosaur and get it all safely out of the ground, so we need to keep working. Besides, we're having too much fun.

We're in Big Sky country, and we can see for miles and miles. Suddenly on the horizon I see thunderheads blowing in, fast. We scramble to get the tools together and the bones covered over with a blue tarp because if rain drenches the freshly exposed bone, it could turn the fossils soft and spongy. And the quarry would become a mud swamp.

We cover the pit just in time and clamber down the hill into our tents and teepees. The storm howls through the site, billowing the tarp. Balls of tumbleweed fly through camp. In one of the storms here this month, hailstones the size of golfballs pelted the ground. In another, the wind blew so hard, it blew the rain sideways and ripped the tops off tents.

In an hour the storm blows over. There's the huge arc of a double rainbow and a fire-red sunset to follow.

YOU'VE GOT TO GET DOWN CLOSE TO THE GROUND, LIKE I'M DOING, TO SEE THE IMPORTANT LITTLE FOSSILS.

The weather puts on one hell of a show in the Hell Creek region. Through these badlands, year round, the wind can blow hard enough to knock your tongue back down your throat. It blows away most everything on the ground until all that's left is dust and rocks. And the bones of dinosaurs. Some of the bones, more than any ever found in any other place known on earth, belong to the king of the dinosaurs: *Tyrannosaurus rex*.

I'm no expert on *T. rex*. I found my first dinosaur bone when I was six, growing up in Montana. Ever since then I've been interested in dinosaurs. But not all dinosaurs. The bone I found was a duckbilled dinosaur, and it is duckbills that I spend most of my time looking for and thinking about. I've found thousands of duckbills since and figured out a lot about how they lived and evolved. And I've found lots of other dinosaurs—horned ones and brontosaurs and little predators and plant eaters. But never a *T. rex*. Then again, all of the people who have ever found a *T. rex* aren't enough to field a baseball team.

I do think *Tyrannosaurus rex* was pretty neat, probably for the same reason you do. It was one humongous, frightening animal. *T. rex* was a genuine monster that's great to fantasize about. It's fun to imagine its life and its world.

And it's exciting to think what will become of the *T. rex* we dig out. A real skeleton of a *T. rex* is something that I can have molded and cast in bronze and put outside the Museum of the Rockies in Bozeman, Montana, where I work. I know a lot of people would come to see it. And while they are at the museum they'd get to see some duckbilled dinosaurs. That makes me happy.

But as a scientist I'm frustrated by *T. rex*. There is only so much science that you can do on one skeleton, or eleven skeletons, which is all we have of *T. rex*. You can find out the systematics—what is related to *T. rex*—by comparing bones. You can look at bones to study the biomechanics of the animal and figure out what it was able to do. But I'm primarily interested in animal

THESE ARE THE BADLANDS
OF EASTERN MONTANA.
KATHY WANKEL FOUND THE
T. REX ON THE HILL IN THE
LEFT FOREGROUND.

ARCTIC
OCEAN

ATLANTIC
OCEAN

PACIFIC
OCEAN

T = TRIASSIC
J = JURASSIC
C = CRETACEOUS
▲ = ALL THREE

AGE OF THE
DINOSAURS

570m. yrs.   220m.   65m. Today

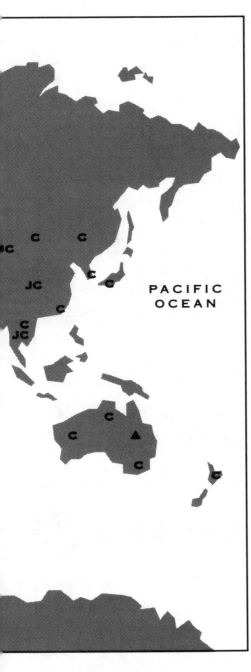

DURING THEIR 160-MILLION-YEAR REIGN, DINOSAURS WERE A WORLDWIDE SUCCESS. TODAY WE FIND THEIR FOSSILS ON EVERY CONTINENT.

PACIFIC OCEAN

behavior and ecology—how dinosaurs acted, what animals they lived with, and in what sort of environment. I can do this for duckbilled dinosaurs because I've found thousands of them, from every stage of development, including embryos still in the eggs and newborns in their nests. I can study their bones and see how fast they grew, study their nests and figure out how they raised their young, study their huge bone beds and figure out how they died. I can even compare their changing kinds over millions of years in one part of western Montana and figure out how changing environments influenced their evolution (some of that research is described in *Digging Dinosaurs*, a book I wrote with James Gorman).

But no one has ever found more than one reasonably complete *T. rex* at a time. No one has ever found the skeleton of a young *T. rex*, or a definite *T. rex* trackway, *T. rex* nest, or *T. rex* egg. And we're lucky to have the eleven *T. rex*es we've got. I and my crew wouldn't have had the chance to dig up the best *T. rex* then known if a rancher hadn't chanced to find it.

Dinosaurs ruled the earth, the whole earth, for 160 million years. Which brings up the question a lot of people, especially aggravated big city reporters, ask: "Why do we have to come to godforsaken places like eastern Montana to find *T. rex*?"

Well, I like eastern Montana. And it is true dinosaurs lived everywhere. We've found enough of them in places like the Arctic and Antarctic and many places in between, from New Jersey to Switzerland to Laos, to know that's so.

Rarely do we find the same kinds of dinosaurs in rock formations from two different times. That's because in the intervening millions of years between the originations of these rock deposits, new dinosaur species evolved and old ones died off. We do know more than three hundred kinds (genera) of dinosaurs now, nearly half discovered in the last twenty years. And the fifty of us who look for dinosaurs—that's all the dinosaur scientists doing fieldwork in the world—find new kinds all the time. There's a new species described every seven weeks on average.

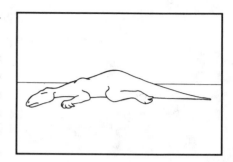

Species of dinosaurs were going extinct all the time over 160 million years. But until the time of *T. rex*, 67 to 65 million years ago, new kinds of dinosaurs were always coming along to replace them. *T. rex* and its contemporaries, duckbills and horned dinosaurs like *Triceratops*, went extinct, too. But after *T. rex* and its neighbors, no dinosaurs came along to replace them.

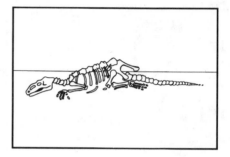

Another reason it's hard to find dinosaurs, even though they lived everywhere, is that there aren't dinosaurs in all the rocks all over the world. The earth is at least 4.5 billion years old. Dinosaurs may be among the most long-lived and successful animals ever to walk the earth. But they didn't come along until 235 million years ago. To find dinosaurs you first have to find the clay or sand or mud—the sediments long since turned to rock—in which dinosaurs died. In many places, the earth was eroding where and when the dinosaurs died. That ground and the dinosaurs in it long ago turned to dust and blew away. For instance, any dinosaurs that lived in highlands will probably never be found, because highland terrain erodes.

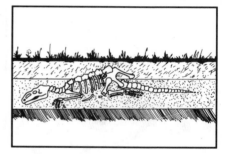

Dinosaur fossils were made only in environments where sand or silt built up—floodplains, stream beds, river valleys, and sand dunes. To make a fossil out of a carcass or a single bone, streams and rivers have to be clogged with dirt and debris and flowing slowly enough to be settling out along the way. A good example today is the Mississippi Delta, where the river has slowed down and is dumping out all the sediment that it's carried downriver.

Eastern Montana at *T. rex*'s time, 67 million to 65 million years ago, was on the eastern slope of the Rocky Mountains, which were then building up to the west. The slope was gradual, and there was enough sediment washing and blowing down from the mountains that layers of what would become rock were slowly building up in the lower elevations. Back then, the sediment covered over the bones of dinosaurs that died near the riverbanks fast enough to keep the bones from eroding. Then, slowly, minerals entered into the spaces inside the porous bones and preserved the bones as fossils.

Now all we have to do is find the rock or the sediments in which the dinosaurs were buried. But a lot of different forces have been shaping the earth before, during, and since the time when dinosaurs died. Besides erosion and deposition, earthquakes and volcanoes and the slow shifting of the plates of the earth's crust have changed the shape of the world and everything in it.

We know, for example, that dinosaurs lived in Indiana. But if you poke around the dirt in Indiana, you're digging in sediments that are 300 million years old (except for surface glacial debris that is much younger than dinosaurs). That's a lot older than the oldest dinosaurs.

But in eastern Montana, sediment of the right age happens to be exposed. So we can walk around out there and literally walk on the ground that the last dinosaurs lived on. However, chunks of this dinosaur time are missing. Each of the striped bands in the hillsides of these badlands represents a different time or weather condition when sediment was deposited. A break could mark a time when nothing was deposited, a time when the environment had shifted to one in which ground stopped building up or even started eroding. We have no way of knowing how long these gaps were, how much time was swept away by erosion.

Or the stripes in the rock could come from continuous sediment buildup that changed in form—from stone made from silt to stone made from sand, marking times when flooding brought in coarser sand grains to the site.

Nowadays, eastern Montana is not depositing anything. It is eroding, fast. The environment is dry and windy, with huge seasonal temperature shifts. Gusting winds, and seasonal freezing and thawing, crumble rock quickly. This extreme weather has been around for only a few thousand years, but it has turned this part

DINOSAUR EXTINCTION

*G*eochemists have spent a lot of time in eastern Montana in the last decade, because it is there that the best evidence of the iridium layer is found. That's the band of the element iridium, rare on earth and more common in rocks from outer space. The iridium-rich layer in rocks in many places around the world, all of them about 65 million years old, is a key piece of evidence for those who think an asteroid or meteor stuck the earth then and produced huge climatic changes that killed off the dinosaurs and many other life forms. Now that there's more evidence of a huge crater made in the Gulf of Mexico at the end of the dinosaur era, many people assume the case is settled—extraterrestial objects killed the dinosaurs.

I can't say I'm much interested in that theory. If an object from space did hit the earth then, there's no direct evidence to link it with the dying off of the dinosaurs at the "K-T boundary" (the end of the Cretaceous period and the beginning of the Tertiary period) 65 million years ago. We don't find any dinosaurs within 100,000 years of the iridium band. In eastern Montana there's no dinosaur skeleton closer than three meters below the K-T boundary line and none above. Maybe dinosaurs lived until then, maybe they lived later. But we don't have the skeletons to prove it either way.

Although there was a sudden pulse of extinctions for many life-forms, including dinosaurs, a lot of other animals, from turtles to mammals, survived. And the sudden extinction doesn't seem to have been worldwide. What evidence we have suggests the Southern Hemisphere might not have suffered from such a catastrophe.

Dinosaur paleontologists aren't expected to look above the K-T line into the Tertiary, though I've searched for fossils from Tertiary stream environments to compare them with the preservation of *T. rex*. The joke among our colleagues is that we'd get nosebleeds if we went up into the Tertiary. It isn't particularly funny or true,

but there aren't many comedians in paleontology.

As for what killed the dinosaurs, there were enough significant, if not sudden, changes in the environment at the time of *T. rex* to account for all dinosaurs' extinction. In the last half of the Cretaceous period (which lasted in all from 144 million to 65 million years ago), the water level rose drastically worldwide. A lot of seaways invaded the continents. In North America, for example, a seaway extended from the Gulf of Mexico clear to the Arctic Ocean until *T. rex*'s time, when it began receding. It allowed for a relatively warm climate way up north. Storms generated in the Gulf of Mexico pumped up into what is now Canada. So we got some tropical climate right along the seaway clear up to Montana. We find crocodile fossils all the way to Canada from this time.

At the end of the Cretaceous those seaways were leaving the continent, and a land bridge had formed, hooking western North America to eastern North America. The air flow and water currents, the temperature regimes and climatic conditions changed, and the climate become more extreme. I'm sure that change had some effect on the dinosaurs. It does seem from some samples that dinosaurs were getting less and less common closer to the K-T boundary.

However, a recent and thorough study by Pete Sheehan of the Milwaukee Public Museum and Dave Fastovsky of the University of Rhode Island showed the opposite—that there was no big drop-off in the diversity of dinosaurs (at the broad level of evolutionary families, not genera or species) in one stretch of time from a few million years before their sudden disappearance to the boundary of their era. It's intriguing information, but I'm not convinced the data relate to mass extinctions. I just don't think we can say for sure how the dinosaurs died or when their decline began.

And the truth is, I don't really care how the dinosaurs died. I'm interested in how they lived. Of course, to figure that out now, I need the bones of dead dinosaurs, the more the better.

of the country into a badland desert. And the ice ages that preceded modern times by tens of thousands of years wiped away far more of the country, carving great valleys far wider than the small streams that now wind across their floors.

Today the badland ground is breaking up, eroding faster than plants can establish themselves. So you don't have any vegetation growing over the rocks and fossils. And not many buildings either, since practically nobody lives out here.

I used to live in New Jersey, where I worked cleaning fossils for Princeton University. There's a lot of dinosaurs there, including the first skeleton of a duckbill ever found in North America (one of my favorites, *Hadrosaurus*). But there you've got to dig under somebody's house, or under a parking lot or a mall, to find a dinosaur. In the badlands, digging for dinosaurs is a lot more convenient and appealing.

How do you actually find a dinosaur? In the past few years, scientists have begun experimenting with sound waves, infrared cameras, magnets, Geiger counters, and a lot of other gadgets to look for dinosaur fossils. So far, none of that high-tech stuff has been proven to work efficiently. And it's too expensive for us, anyway.

To find a dinosaur, I just walk up and down the bare hills with my head down and stop when I come to something that looks like a dinosaur bone. Sometimes I crawl around on my hands and knees to find little bones. I know that sounds too simple, but it's pretty much the way dinosaur hunters have done it for 150 years. Dinosaurs could be anywhere in badlands like those in eastern Montana. So if you're looking for a fossil, you pretty much just have to go up and down through the hills. It takes a lot of persistence. You could look over the side of one hill and never find anything. And if you didn't go on the other side of the hill, you'd miss the find of your life. So I try to hit every spot.

If I have success where no one else has, maybe it's because I'm more persistent. Or maybe I'm just lucky. I've found fossils when I was just trying to dig a hole for

DURING THE LAST DINOSAUR PERIOD, THE CRETACEOUS, A HUGE SHALLOW SEAWAY MOVED IN AND OUT ACROSS THE INTERIOR OF NORTH AMERICA.

TOP: SEVENTY-FIVE MILLION YEARS AGO THE SEAWAY CAME CLOSE TO THE ERUPTING VOLCANOES OF THE GROWING ROCKY MOUNTAINS.

BOTTOM: BY *T. REX*'S TIME, 65 MILLION YEARS AGO, THE SEAWAY HAD CONTRACTED, EXPOSING ALL OF PRESENT-DAY MONTANA.

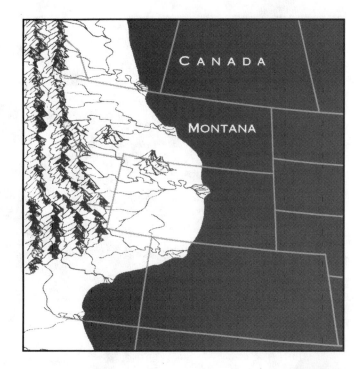

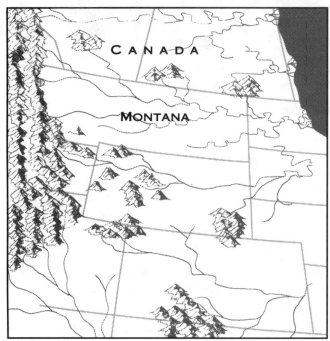

my tent stake. Nice accidents happen to other fossil hunters too. My friend Phil Currie, a dinosaur paleontologist in Canada and an expert on *T. rex*, once found the skull of a predatory dinosaur when his camera case rolled down a hill and came to rest on top of it.

None of us can get to every promising spot. And what's exposed changes quickly in the badlands, so today's promising fossil find will be dust in twenty-five years if it isn't excavated and preserved.

What you find depends also on what you are looking for. Someone else might be looking for big specimens and so might stand on a hill and look around to scan for big bones. I'm interested in every fossil, including the little ones. So I cover a lot of ground.

ANIMALS GATHER BY WATER, AS THESE WILDEBEEST AND ZEBRA DO IN AMBOSELI, KENYA. THEY DIE THERE, TOO, AND HAVE A CHANCE OF BEING COVERED OVER BY SEDIMENT AND PRESERVED AS FOSSILS.

But I do most of my looking in western Montana, the land I grew up in and know best. If I dig a dinosaur someplace else, like the *T. rex* in eastern Montana, it's because someone else found it and asked me to dig it up. That's just how it was with the *T. rex* we dug up.

The next question people ask me is usually, "How come *T. rex* is found only around Montana and South Dakota?" Most, but not all, of the few *T. rex*es we know do come from eastern Montana and western South Dakota. The answer to that one has to do with both the location and the age of the rocks. Dinosaurs lived all over the world, but not the same dinosaurs. *T. rex* seems to have made its home in western North America—we've found its remains in Alberta, Canada, as well as Montana, Wyoming, Colorado, and South Dakota. So far, we haven't found the same animal on any other continents. But there is a nearly identical creature, named *Tarbosaurus*, which lived near to *T. rex*'s time in Mongolia. That's not so surprising because close Asian cousins to a lot of the dinosaurs of western North America lived at the end of dinosaur days. The two continents were joined from time to time in this period, and populations of dinosaurs could spread across the Arctic to new territories.

The other reason we can find *T. rex* in eastern Montana has to do with time. Each rock formation we explore for dinosaurs is a window into one specific relatively brief period of time. In the Hell Creek Formation that time is a few million years, ending 65 million years ago, the time of *T. rex*.

We know the age of those rocks better than almost any others from dinosaur time. I can't figure the age of rocks precisely. I'm a paleontologist—I study fossils. But some geologists and physicists have been able to date those rocks. To get a good "absolute" date for the age of rock you need a certain kind of rock. The kind that is most likely to contain dinosaurs is sedimentary rock. However, for geochemical dating you need rock formed suddenly by volcanoes, at one single time. That's igneous rock.

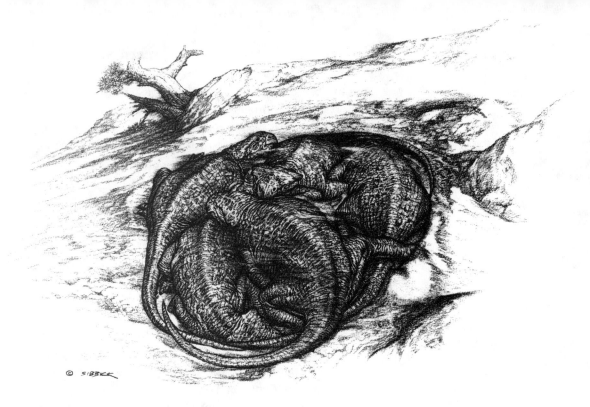

In the Hell Creek Formation we're lucky to have not just sedimentary rock but igneous rock, in the form of ash beds from volcanic eruptions. In that ash are tiny, but detectable, radioactive isotopes of elements. Over millions of years some of those elements decay into other elements, at a constant rate. For instance, radioactive potassium eventually breaks down into argon, an inert gas. Geochemists know the rate of that change. If they measure the amount of potassium in the volcanic ash versus the amount of argon, they can tell, give or take a hundred thousand years, how much the composition of isotopes has changed. When we know how much change has occurred in how long a time, we can figure when that rock and the animals in it were made. We call this radiometric dating.

Scientists know the age of many forms of extinct life precisely, no matter what some creationists say. The idea that scientific dating methods could be off by billions of years, and that humans ran around with dinosaurs, is not true. But that hasn't stopped many

SOMETIMES WONDERFUL FOSSILS ARE CREATED IN ARID CONDITIONS, AS WHEN THESE BABY ARMORED DINOSAURS IN THE GOBI DESERT WERE SUFFOCATED IN A COLLAPSED SAND DUNE. CHINESE AND CANADIAN PALEONTOLOGISTS FOUND THE SKELETONS OF THESE BABIES PERFECTLY PRESERVED NEARLY 80 MILLION YEARS AFTER DEATH.

people from believing those things. Just the same, it's important to keep our information straight and not confuse the fact of evolution with ideas from biblical authors and interpreters.

Though it isn't full of errors, radiometric dating of rocks is a difficult and time-consuming business. You don't just run around and say, "OK, I've found some really nice ash beds here. I'm gonna take home a piece of rock and date it."

Usually, good chemical dating isn't available to us, and we rely on other less precise, but still pretty useful indications for the age of fossils. Paleontologists have compiled a pretty thorough collection of small mammals, pollen spores, and tiny marine creatures from various dinosaur times. Find any of these near a layer of rock with dinosaur bones in it, and you can get a pretty good fix on the age of the rock. And dinosaurs themselves can help date the rocks. We know something of which dinosaurs lived at which time. So if you find *T. rex* and *Triceratops* bones, it's a pretty strong indication the rock they are in is 65 to 67 million years old.

If I wanted to find a dinosaur bone of a certain age, I would first look on a geological map for sediments of that age, then go to that place and hope it is all busted up into badlands that make it easy to follow the age of one layer of sediment. I'm relying on data from geologists, who have mapped this land pretty thoroughly. And geologists know about changes in the terrain, like the upthrust of mountains, and when they happened. I've learned a bit of geology myself, enough to figure out what the local environment looked like in dinosaur times (see Chapter 4).

For instance, the yellowish rock is sandstone left by streams, and the grayish rock is mudstone or siltstone made on the floodplain when the streams flooded. That is simple enough, and you find fossils in both stream channel and floodplain environments. What is harder is reading the breaks in the stripes of sediment where the ancient stream channels ran, and finding one layer of time across the crumbled and jumbled layercake of the parched badlands of eastern Montana.

But none of this information can give you a good idea of what time you are in when you walk through the badlands. Channel and floodplain deposits like those you walk in here look pretty much the same, no matter how old they are.

So, knowing all this, how come we have only eleven incomplete skeletons to show for a century of digging? With good people looking and lots of information about the time and place of *T. rex*, you would think we might find more of it. This was, after all, a huge animal, and its kind seems to have survived for several million years across the West.

Truth is, dinosaur skeletons are pretty scarce, period. We've got only about 2,100 of them in all the world's museums. A lot of dinosaurs we know only from a single tooth or chip of bone. It takes special conditions to make a fossil, and a lot of luck to find it. Even if you've found the rock of the right age, with a *T. rex* fossil in it, you've got to come on it at the right time. Rocks, and the fossils in them, erode fast in the badlands. One season a bone is freshly exposed. The next it may be dust, "exploded" as we say. The best fossil specimens, like our *T. rex*, are those where just a tip of bone is exposed. Then you can dig down a few feet and get out the rest of it before the elements get at it. And it was lucky that our *T. rex* was in sandstone, not mudstone. Mudstone has a lot of volcanic ash in it, stuff called bentonite. Bentonite is used to plug up dams, because it swells up in water. In the badlands, when bentonite gets wet it swells up just the same way. Then it shrinks

THE BEST FOSSIL FINDS ARE OFTEN THOSE WHERE ONLY A BIT OF BONE STICKS OUT ABOVE THE SURFACE. IN THOSE CASES MOST OF THE BONE IS STILL SHELTERED FROM EROSION. THAT'S HOW IT WAS WITH KATHY WANKEL'S *T. REX*.

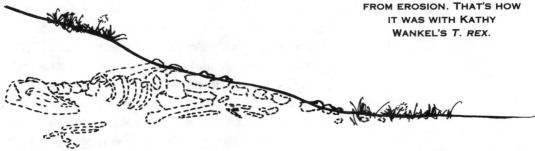

during the long dry spells. When it shrinks it cracks all over, making the ground into a layer of cracked crust over dust, stuff we call "popcorn." You can just imagine what the popcorn cracking does to bones. If a *T. rex* died and was fossilized in bentonite, it could end up in a million pieces.

Yet another reason it's so hard to find a dead *T. rex* is that there probably weren't that many live ones. *T. rex* was a meat eater, and fossils of meat eaters are always a lot harder to come by than those of the animals they ate. That's because there are always a lot more plant eaters. Think about the African savannah today. For every pride of lions, wildebeests roam in the tens and hundreds of thousands. The same probably was true for dinosaur predators. It would help explain why 85 percent of all dinosaur carnivores are known from five or fewer specimens. And nearly half the species of meat-eating dinosaurs have only one specimen to their name.

T. rex has been such a familiar, popular animal for so long that it's difficult to believe that we hardly know it. But the truth is, we're lucky to have any proof of it at all.

In the space of two months, the two most complete *T. rex*es ever found were unearthed (ours and "Sue" in South Dakota). That's luck. But getting a *T. rex* out of the ground takes skill, and still more luck.

We have more electrical equipment than the early dinosaur diggers had, from little air hammers to big pneumatic drills and backhoes. Dave Gillette, the Utah state paleontologist, has been experimenting with remote sensing equipment to find dinosaurs by sound waves, magnets, and infrared rays. For a long time dinosaur bones have been detected with Geiger counters, since many fossils are "hot" with uranium.

But mostly we dig dinosaurs exactly the same way as the early dinosaur diggers in the American West did one hundred years ago. We still use picks and shovels to get down to the bone layer, still wrap the bones in burlap and plaster, still brace them with wood, and still haul them out to the laboratory for cleaning.

We go more slowly, though, than the early collectors. They were spurred on by competition among the big eastern museums like Yale University's Peabody Museum of Natural History, the Carnegie Museum of Natural History, and the American Museum of Natural History, all of which wanted lots of stuff to display in the museum right away. So the prospectors used dynamite to blow the tops of hills, risking the fossils. They stole fossils and shot at each other. And whatever they found that looked big enough to display, they'd bring back to the museum. There several preparators would clean the fossil up quickly, and a scientist would write a short paper describing it as a dinosaur and invent a name for it. Then they'd put it together on steel rods, stick it out on the display floor, and go look for more.

Nowadays we're trying to answer several biological questions when we dig up dinosaurs. We do take things like this *T. rex* back to the museum, and a cast of it will make a great showpiece. But we're also trying to learn something about the animal. So we dig up *T. rex* more slowly. We take notes about the kind of sediment it is in. We map it to see how the bones lie in the ground, so that we can try to figure out the circumstances of its death and burial. We sift through all the dirt and rock around the bones, in the field and in the lab, looking for other dinosaur bones, mammal and reptile fossils, fossil leaves, and pollen spores—anything that will tell us what kind of environment this animal lived in. And died in.

The study of what happened to animals after they died and how they were buried is called taphonomy. A German scientist pioneered it back in the 1920s, and Peter Dodson of the University of Pennsylvania was the first to apply it systematically to dinosaurs, twenty years ago. He studied the way dinosaur bones in Dinosaur Provincial Park in Alberta, Canada, got moved around by ancient stream channels.

It's very difficult to tell for sure how an animal died because you very rarely find it preserved just as it looked in the moment of death. That's true even for a

ABOVE: THIS IS ONE OF THE MOST REMARKABLE OF ALL FOSSIL FINDS, FROM THE STATE MUSEUM IN ULAN BATAAR, MONGOLIA, SHOWING A CARNIVORE, *VELOCIRAPTOR*, LOCKED IN BATTLE WITH A *PROTOCERATOPS*. BOTH WERE APPARENTLY KILLED SUDDENLY BY A GOBI DESERT SANDSTORM.

RIGHT: ARTIST ROBERT WALTERS DEPICTED THEIR BATTLE IN THIS PAINTING.

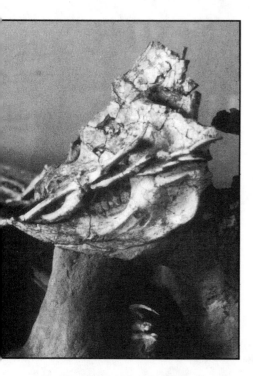

recently dead animal, unless of course you find a bullet hole in its skull.

There are exceptions. One set of dinosaur fossils from Mongolia shows a small predator and a horned dinosaur wrapped around each other. It seems some tragedy befell both of them before one of them could kill the other. And a huddle of young armored dinosaurs found in a sand dune in Inner Mongolia a few years ago probably died just as they lay from the collapse of the sand dune. And maybe we'll find a duckbill with a *T. rex* tooth buried in its skull. We did just discover a duckbill leg bone chomped on by a close ancestor to *T. rex*, a dinosaur we're getting ready to name. A piece of the carnivore's tooth was still in the tailbone.

The excavation of our *T. rex* will never tell us how it died. But we will discover a lot about how it looked and moved, and something about the conditions that buried and preserved it and the environment near the end of dinosaur time.

TOP: *T. REX* AS A SCAVENGER,
EATING A *TRICERATOPS*.
DRAWING BY DOUGLAS HENDERSON.

BOTTOM: *T. REX* AS A PREDATOR,
HERE ATTACKING A *HADROSAUR*.
PAINTING BY DAVID PETERS.

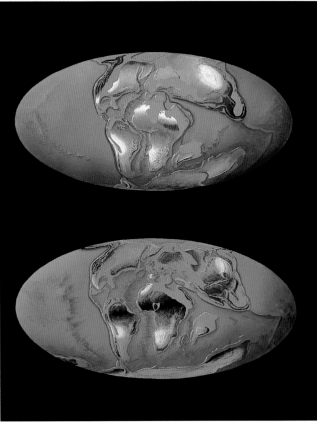

ABOVE: *T. REX* ATTACKS *TRICERATOPS* IN THIS FAMOUS CHARLES R. KNIGHT PAINTING FROM THE 1930'S. THE *T. REX* MAY BE A PORKER BUT KNIGHT DID HIM LEANING FORWARD WITH HIS TAIL OFF THE GROUND. FIELD MUSEUM OF NATURAL HISTORY, CHICAGO.

TOP RIGHT: *T. REX* LOOKS A LOT SWIFTER IN THIS MODERN PAINTING, COPYRIGHT GREG PAUL, 1988.

RIGHT: *T. REX* AS A COOPERATIVE HUNTER. HERE *T. REX* CIRCLES A HERD OF *TRICERATOPS*. I DON'T BELIEVE THIS SCENE FOR A LOT OF REASONS, BUT IT IS A NICE PAINTING BY MARK HALLETT.

LEFT: MUCH OF WESTERN NORTH AMERICA WAS UNDERWATER IN THE HIGH WATER DAYS OF THE MID-CRETACEOUS, 94 MILLION YEARS AGO. BY *T. REX*'S TIME, 67 MILLION TO 65 MILLION YEARS AGO, THE SEAWAY HAD SHRUNK A LOT. MAPS BY CHRIS SCOTESE.

NEXT PAGE: *T. REX* IN ITS ENVIRONMENT AS DRAWN BY DOUGLAS HENDERSON.

SOME OF *T. REX*'S
CARNIVOROUS COUSINS:

ABOVE LEFT: *T. REX*'S CLOSEST
RELATIVE AND NEAR LOOK-ALIKE,
TARBOSAURUS FROM MONGOLIA.
DRAWING BY DOUGLAS
HENDERSON.

ABOVE RIGHT: A SMALLER AND
SEVERAL MILLION YEARS MORE
ANCIENT RELATIVE,
ALBERTOSAURUS. PAINTING BY
GREG PAUL.

FAR RIGHT: PERHAPS *T. REX*'S
DIRECT ANCESTOR,
DASPLETOSAURUS. PAINTED BY
GREG PAUL.

RIGHT: A COMPACT
CONTEMPORARY RECENTLY NAMED,
NANOTYRANNUS. PAINTING BY
BRIAN FRANCZAK.

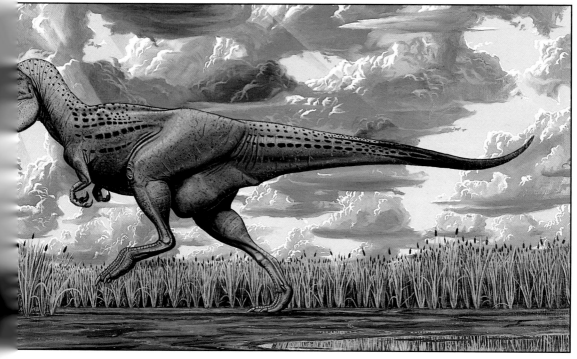

BELOW: THE CREW'S AT WORK ON *T. REX* HERE, HAVING ALREADY TAKEN DOWN TONS OF ROCK FROM WANKEL'S ISLAND.

LEFT: HERE'S A BUNCH OF US (I'M THE ONE IN THE RED VEST) LOOKING AT THE NECK OF THE EXPOSED *T. REX* AND FIGURING OUT HOW WE'RE GOING TO GET SOMETHING THIS BIG OUT OF THE GROUND.

BOTTOM: SUNSET OVER WANKEL'S ISLAND. PHOTOGRAPHY BY BRUCE SELYEM, MUSEUM OF THE ROCKIES.

CHAPTER 2

THE DIG

THERE MIGHT BE MORE dinosaurs than people in McCone and Garfield counties in eastern Montana. In fact, they're among the least populated counties in the continental United States. That's not surprising. This is badlands country, and the weather's pretty strange. It gets up to 120 degrees in summer, down to -50 degrees in winter. I think it's beautiful, but then I grew up walking around Montana badlands.

Kathy Wankel is one of those few people who live in McCone County. She and her husband, Tom, have a ranch about sixty miles outside the town of Jordan. The closest neighbor is thirty miles away. When I met them, their three kids went to school twenty minutes down the road from their house. Well, it isn't exactly a school. It's a mobile home. Half is the teacher's house, half is the school. The Wankel kids and two others were the whole student body, kindergarten through junior high.

The Wankels work hard. But they do take Labor Day off. On Labor Day afternoon in 1988, Kathy was fishing with her family in an arm of the Fort Peck Reservoir in the Charles M. Russell Wildlife Refuge. It's a big stretch of badlands about thirty miles from Kathy's house. Except for the cattle that graze on what little grows there, the only animals you're likely to see there are white pelicans and blue herons, a few coyotes, badgers, and a zillion rabbits and ground squirrels. Chances are, even on a holiday, you won't run into any people.

But there's a good chance you'll come across a dinosaur. The hills and gulleys there are made of sand and mud from the time of the last dinosaurs. And the bones of those dinosaurs are scattered throughout

those sediments exposed in the badlands near Jordan.

The Wankels know that well. They pulled their boat off the lake by a hilly sandstone island. Kathy and her kids began strolling up the hill. The slanting sun lit on a piece of bone, and Kathy caught the telltale glint. What she saw was a little ridge of bone sticking out of the ground. She started digging with a pencil to see how big a piece of bone she'd found. The more she dug, the larger the bone seemed to be. As Kathy recalls, soon she was just shaking all over, saying, "I've made a megafind, a megafind." She had a feeling this was something big, maybe a *T. rex*. In her experience, it was something big. And it turned out to be a megafind by anyone's standards.

Kathy had the good sense to stop digging and to bring the bones she'd found to a museum. Fortunately for me, she went to the Museum of the Rockies in Bozeman, Montana, which has the biggest paleontology collection in the state. That's where I work. At Thanksgiving, Kathy and Tom drove that bone three hundred miles across the state to show it to someone at our museum.

People are bringing us bones all the time, and we're

THE CHARLES M. RUSSELL WILDLIFE REFUGE. THE FORT PECK RESERVOIR HAS SHRUNK TO THE LEFT, EXPOSING THE LAND BELOW WANKEL'S ISLAND, THE HILL TO THE RIGHT WITH THE WHITE PATCHES OF PLASTER ON TOP.

T. REX DISCOVERER KATHY WANKEL, HER HUSBAND, TOM, AND THEIR KIDS CAME OUT TO HELP US EXCAVATE.

happy to see them all, especially if it means folks haven't dug any farther and have left that job to us. I know it's hard to resist the urge to dig up buried treasure once you've found it, but there's a lot of information lost and fossils ruined by amateur bone-hunters who insist on excavating fossils themselves.

Kathy showed the few bones she'd dug to me and Pat Leiggi, my longtime field crew chief. Right away I could tell Kathy had found the shoulder and arm of *T. rex*, even though only once had anyone found those parts of the animal before. I was pretty excited myself, though I'm so low keyed, folks say they can't tell the difference on me. And maybe I didn't hop up and down because I realized the amount of work that would be involved to get the rest of the animal, in the off chance more of *T. rex* was out there.

I knew what Kathy had brought me belonged to the arm of a *T. rex* because I had some important clues. The badlands Kathy was looking in belonged to the Hell Creek Formation. That's rock from *T. rex*'s time, 67 to 65 million years ago. And the Hell Creek Formation is the source of most of the world's good *T. rex* skeletons.

Also, the bones Kathy found were hollow. Like birds, meat-eating dinosaurs had air cavities in their bones. But by their shape and size and age these bones could belong to only one animal—*T. rex*.

We told the Wankels what we thought they'd found and asked them to keep it under their hat. Tom Wankel said he'd have to get a bigger hat. But they kept their big secret well, for more than a year.

I didn't have time to get out to the site right away, but Pat Leiggi did, as soon as the weather got warm. That wasn't until May of 1989. Meanwhile I gave the *T. rex* upper arm bone to two former Museum of the Rockies workers who know a lot about how to interpret dinosaur anatomy and movement from bones: Matt Smith, a sculptor and preparator, and Ken Carpenter, an exhibit designer and researcher.

Pat took Ken and Matt with him when he went out to the Wankels' site in May of 1989. By then it wasn't an

island any more. The reservoir had receded far enough that the *T. rex* site was now a flat-topped hill a quarter-mile inland from the shore. They scraped around a bit with pickaxes in soft sandstone at the top of the hill, and Matt saw something he thought was the boot of a pubis, the broad bottom of one of the bones of the pelvis. Pat told him to cover it up. When they got back to the museum Pat said this was something we should go after, and I asked him to head up the dig the next year. I needed Pat to oversee digs already scheduled for that summer. And I knew he, not I, would be the best one to run a *T. rex* excavation. With a laboratory to manage, papers to write, and prospecting for new sites on my schedule, I knew I'd never be able to organize it. Besides, nobody knows more about how to handle the logistics of a dig than Pat.

Pat's still a young guy, just hitting forty, but we go way back. He was my first employee, cleaning fossils for me

ABOVE: MATT SMITH, SHELLEY MCKAMEY, AND BOB HARMON ARE DOING SOME PRELIMINARY EXCAVATION AND MAPPING OF *T. REX* BONES IN 1989. THE STRING GRID IS USED TO PLOT WHERE THE BONES WERE FOUND.

RIGHT: ALLISON GENTRY IS APPLYING A PLASTER JACKET OVER THE PARTIALLY EXPOSED RIBS OF KATHY WANKEL'S *T. REX* IN THE FALL OF 1989 TO PROTECT THEM UNTIL WE RETURN TO EXCAVATE THE WHOLE SKELETON THE NEXT JUNE.

when I was a researcher at Princeton University in 1980. (I started at Princeton as a preparator myself in 1975.) Then Pat worked as a preparator for Ken Carpenter at the Academy of Natural Sciences in Philadelphia. Now Pat works for me at the Museum of the Rockies. And for a while until he became a preparator at the Denver Museum of Natural History, Ken Carpenter worked for Pat at the Museum of the Rockies. It's a small world.

In September of 1989, as soon as the summer's work was over and before the ground froze, Pat went back to the *T. rex* site with Matt Smith; Carrie Ancell, the museum's senior preparator; and Bob Harmon, museum preparator and jack-of-all-trades. They had to walk about two miles to get there, carrying generator, jackhammer, water, and food. They had less than a week to explore, but right away they knew they were onto something big. They found two-thirds of the pubis, a skull, and a right leg beside it. They really got excited when Pat probed around and hit a vertebra. He went to the side of it and came on another. He scraped

at the other side and found one more. When Pat came on those vertebrae, he says he felt "like a kid in a candy store." His find meant these backbones were still articulated, buried in place, and that there was a good chance a whole *T. rex* lay in this hill.

Pat knew he didn't have the time and the work force to get this *T. rex* out right away. So they covered the bones with a thin plastered burlap cast, trenched around them to drain off water in case of rain, covered the site with dirt for the winter, and asked Jim Alphonso, assistant manager of the Charles M. Russell Wildlife Refuge, to look after it.

To excavate the *T. rex* the next spring, Pat had to begin that winter by getting the necessary permission from Jim's bosses, the owners of the *T. rex* site. Wankel's Island (as we call the site) and the land and water for miles around it are part of the Charles M. Russell Wildlife Refuge, the CMR, a federally owned, Department of Interior–administered hunk of badlands and flatland. The CMR is open for grazing cattle and recreational use. You need what's called an antiquities permit to excavate on federal land, so early in 1990 Pat

I FOUND BOB HARMON DOING WHAT I DO, WALKING AROUND MONTANA LOOKING FOR DINOSAURS. HE'S PROVED AN EXPERT FOSSIL PREPARATOR AND FIELD WORKER.

went to see Roy Snyder and Phil Sheffield of the Army Corps of Engineers at Fort Peck, about seventy miles away from Wankel's Island. They talked it over with John Foster and Bill Haglan of the CMR, and Pat said they were all "unbelievably cooperative. They asked what I needed. I said, 'A road would be nice,' and they said they'd make us one."

For months after that Pat talked to Jim Alphonso about the road. It was a good couple of miles across rocky ground and a nearly dry arm of the Fort Peck Reservoir to the hill where *T. rex* lay. It wasn't easy to figure out where to put the road. Pat finally came up with the idea of going straight through a dry creek bed.

With a bulldozer the Corps cut us a great road, across a cow pasture and a little washout below the reservoir (the CMR put a combination lock on the fence), then winding along the badland hillsides right up to the site. Pat set aside June and readied a crew and equipment to get *T. rex* out inside of a month. At last, we were ready to dig.

When Pat allowed a month to dig the whole *T. rex* out, he had to consider not just the mechanics of getting the *T. rex* out of the ground, but the availability of the Army Corps of Engineers and its equipment and all the other volunteers who were helping us out, media, money, and weather.

But thanks to good planning from Pat and good weather, the excavation went like clockwork. Pat, Bob, and Allison Gentry, another of the museum's top preparators, arrived on June 4 to pull off the protective winter jacket and the brush they'd put over it to hide the site. A few days later the rest of us turned up and started our six-day a week schedule (Sundays were reserved for the hour drive to Jordan for a visit to a shower, laundry, and supermarket).

Pat had put a dozen of our top diggers on the *T. rex* excavation. It was the best, most experienced field crew in the world.

Pat's crew chief for the *T. rex* dig was Bob Harmon. I didn't recruit Bob, I captured him. We were prospect-

DAVE VARRICCHIO, MY DOCTORAL STUDENT, IS ALSO HANDY WITH A JACKHAMMER.

ing at one of our sites on the Two Medicine River in western Montana one summer morning when we came on someone out looking for fossils. I wanted to know what the heck he was doing there, but I was pretty respectful. Bob's a serious-looking guy and he owns a serious-looking gun.

Turns out Bob was just driving and walking around looking for fossils. And he was pretty darn good at finding them. I put him to work digging for us, and he's found all kinds of great stuff for the museum, from dinosaur skeletons to nests. Bob's strong too, and he knows how to handle a six-foot prybar or a sixty-five-pound jackhammer. We had others from the museum out working too: Carrie, Allison, doctoral student Dave Varricchio, Matt, Bea Taylor (a trustee of the museum who is also one of our best preparators and a decade-long fieldworker), and the museum's public relations director, Shelley McKamey. We also used some of the top volunteers we've had from past digs, many of whom are not scientists. Brad McMullen, for example, is a physician who had worked in the field the year before and was so good we asked him to take a working vacation and help us with *T. rex*.

We have to rely on volunteer help. We can't afford to pay for the work force we use, and our entire budget for excavating *T. rex* was only $5,000. We get lots of requests every summer from people who want to come and spend a few days digging dinosaurs. It's hard to tell from a letter who's going to be good at it. So I throw all the letters out. I'm not so great about answering letters anyway. If they write back, and let me know they wrote before, then I pay attention. I figure they must really be interested to write twice, or more.

TOP: BEA TAYLOR AND BOB HARMON PLASTER OVER THE BOTTOM OF THE *T. REX* HIP BUNDLE.

BOTTOM: WE'VE GOT MOST OF THE WANKEL *T. REX* SHOWING NOW. I'M SITTING IN THE FOREGROUND TRYING TO FIGURE OUT HOW I CAN SEPARATE THE LEG AND SKULL BONES, WHICH HAVE WASHED TOGETHER, WITHOUT HARMING THE FOSSILS.

Some of those who come out turn out to be great diggers. They learn the techniques—like scraping gently around the bone with the awl, even a toothbrush sometimes, never digging into the ground, and squeezing on the "Vinac," the clear liquid that hardens the bone as it's exposed. Or how to "jacket" a bone. You've got to mix the plaster to the right consistency so it spreads easily onto the bandages and over the paper towels separating it from the bone and matrix and then dries fast and hard. You've got to "pedestal" the bone, digging way under it before you turn it over to jacket the bottom side.

It's not brain surgery, but you'd be surprised how many people screw up at digging fossils. You need a soft touch and good eye. Most of all you've got to be patient. You never know what it is you've got, and you've got only one chance at digging it out of the ground. Even with the best of intentions, untrained amateurs can damage an important specimen. We would never have known about this *T. rex* if Kathy Wankel hadn't brought us the arm bones. But those bones would have been in better shape today if she'd left them in the ground, and let us take them out. (Of course it might have been harder then to convince us to come out and look at them.) And Kathy was as careful as any amateur could be in her digging.

Some professional paleontologists never get the hang of digging fossils either. Roy Chapman Andrews, the great American Museum of Natural History adventurer of the 1920s, was so bad at digging bones that researchers at the museum still refer to a damaged specimen as "RCA'ed" in his honor.

To excavate something as rare and important as a complete *T. rex* we wanted only our best diggers. We also knew we needed the job done pretty quickly. Our resources were limited, and the crew already had a full summer of fieldwork ahead, digging up sites I and others had prospected the year before. I had some potential duckbilled and horned dinosaur sites in the Two Medicine Formation of western Montana I was itching to look at as early as possible that field season.

Our crew had to be more than talented. They needed to be in good physical shape. Digging dinosaurs can be back breaking, literally. By and large we don't have many injuries on the job. But Pat has chronic back trouble from lifting rocks to get at bones in the field. Vicki Clouse, one of our student helpers on the *T. rex* dig, broke a bone in her foot when something heavy landed on it. She had to quit work. In the past, dinosaur diggers have even been killed, crushed by falling rocks or equipment. Aside from Vicki's injury, the worst incident we've had at a site was a rattlesnake bite in the ankle that laid up Carrie in the summer of 1992. Still, digging dinosaurs is safer than the long drives between sites we often make on back roads at late hours. A crash on one of those drives is what killed my first excavation partner and best friend, Bob Makela, several years ago.

Pat didn't dig a single bone out at the *T. rex* site himself. Not that he was protecting his bad back—he ran the jackhammer and pushed the crowbar enough to screw up his back worse than ever. But he was busy supervising the rest of the crew and all the equipment, devising seat-of-the-pants strategies for taking each group of bones out as they were uncovered.

Good as the road was, it was only one lane, and rain, wind, and all the vehicle traffic pretty soon made for teeth-rattling ruts and rough spots. In the rain the crushed shale got so slippery, four-wheel-drive vehicles just slid right off.

The condition of the road, the cooperation of the folks in the area, and the padlocked fence were enough to keep out most uninvited visitors. We did get one nervy guy who was driving cross-country from Vermont, heard about our dig, and kept asking people around Jordan until one of them told him where to go. Instead of tossing him out, we showed him around the dig. I figured if it meant that much to him to see it, he ought to have that chance.

We did have a lot of media people to contend with, and that was sometimes Pat's biggest headache. We

WE HAD TO WORK OUR WAY AROUND THE TELEVISION CREWS. HERE PAT LEIGGI IS TRYING TO KEEP A SAFE DISTANCE FROM A NOVA TEAM.

wanted the press there. It's good publicity for the museum and the state, and showing people what we do is the educational message we want to get out. And you can't play favorites once you invite the press in. But doing that without turning it into a media circus isn't easy. Shelley controlled the free-for-all by scheduling press days, as well as special viewing days for folks from the area, and for the brass of the army, which did so much to help us.

We knew digging up the best *T. rex* yet would attract the press, but we were still surprised by their frenzy. One national network show wanted to send a remote truck to do a live transmission from the site. When Shelley explained that their satellite truck would probably never make it over our road, and certainly not with its equipment working, they had another idea. Could I fly out to Los Angeles by private plane in time to appear live on their morning show? Not only would that have meant getting up long before dawn, something I had no intention of doing, but the trip probably would have cost more than our dig.

As it was, a network show and three different American television documentary crews and one from Japan came to tape the dig. Shelley had to schedule them each for different days. One of them even faked our trucking the *T. rex* away since they couldn't wait around for the real thing. And more than once a crew stepped on a researcher or a bone, or stumbled onto one of us using the area reserved as an open-air latrine. These events sent Pat's blood pressure soaring. Dozens of reporters and photographers came and went (not including the museum's own photographer, Bruce Selyem, who documented much of the dig)—even one guy who claimed to be from the Associated Press, a great surprise to the guy who *was* sent by the Associated Press.

So Pat had to figure on all these diversions when planning his dig schedule. He also had to plan, as best anyone can, for the summer weather in eastern Montana. It never rains much in this country, but the last few years had been real drought times. Still, with the temperature well over 100 degrees most days, the thunderheads would build up, and by day's end a fierce storm was sure to roll by. Usually you saw only rain off in the distance, along with lightning strikes and the occasional rainbow. Together with a fiery sunset it made for a heck of a show every night. But a couple of times the storms dropped on camp. The worst one packed hail and winds of seventy miles an hour that ripped Pat's tent to shreds and tore the weighted tarp off the quarry. Fortunately it didn't rain enough to soak the bones, and the crew had enough warning to trench the site, so the water ran off quickly.

Every workday at the dig, except for flash storms and media blitzes, the schedule was the same. We'd be up and out of our tents and teepees by 7:00 A.M. Not much time was spent on personal grooming. It was eat and run from the cook tent to start working before it got too hot.

Not that we were smart enough to stop working in the middle of the day. The crew took sandwiches along

WITH PRY BARS AND SORE BACKS, THE CREW IS HACKING AWAY TONS OF SANDSTONE OVERBURDEN TO GET NEAR THE WANKEL *T. REX* FOSSILS.

and usually kept working until past 6:00 P.M. Then it was dinner, beer, horseshoes, and conversation. I was most interested in the beer and the horseshoes.

In the daytime I was rarely around to see what was doing. Instead, I was off walking around looking for more fossils, making maps and notes. Prospecting is not always productive, and sometimes it's dangerous. Once, looking out for fossils and not where I was going, I caught my heel going down a slope and tumbled and slid a good fifty feet. Less than a mile from the site I did come on two *Triceratops* horns, but the rest of the skull wasn't much to look at. We also found a champsosaur (a four-foot long aquatic lizard of *T. rex*'s time) worth collecting.

Meanwhile, Pat and his crew were removing at least fifty tons, and probably more like eighty tons, of sandstone, using ten-, thirty-, and sixty-five-pound jackhammers, along with Bob's crowbar to peel away big sandstone blocks. Everyone pitched in to roll these boulders down the hill. We even got help from Bob Sloan, a mammal paleontologist from the University of Minnesota who's been working the Hell Creek Formation for decades. The scraps were shoveled into buckets and hauled to the edge, then tossed over.

The crew had to take down most of the hillside to get at *T. rex*. By the time they were done excavating, the pit itself was about forty-five feet across and twenty-seven feet into the hill, and they'd dug down about twelve feet into the hill. It took about twelve days of hard work to get all that rock out.

As Pat's crew got through this overburden and close to the layer of sediment where they'd found parts of the skeleton the year before, they switched to smaller and smaller tools—picks and hammers, then awls and trowels, and finally small brushes.

As the brown bone was exposed, it was soaked with squirts from a detergent bottle filled with Vinac. The smelly liquid not only hardens the bone, it makes the bone shiny. Often, even after it has been treated with Vinac, bone will be soft, spongy, and crumbly. We try to handle it as little as possible in the field, since even the moisture from our hands can make it deteriorate. The matrix, the rock that surrounds it, can present just the opposite problem. Ironstone—or in the case of Kathy's *T. rex*, hard sandstone—can be hell to dig through. Fortunately the sandstone immediately around the bones was not as dense as in the overburden. And this *T. rex*'s bones were in pretty good shape, with the exception of the exposed arm and

THE SKELETON OF A NEWLY DEAD *T. REX* AS IT LAY STILL COVERED WITH SKIN AND FLESH.

MUSCLES CONTRACTING IN DEATH RIGOR PULLED THE HEAD BACK BEFORE ROTTING AWAY, ALONG WITH SKIN.

HERE IS THE WANKEL *T. REX*
EXPOSED. ITS NECK IS TO
THE LEFT, THE SKULL IN THE
CENTER BELOW THE HIPS.
SEE THE DIAGRAM BELOW
AND ON THE FOLLOWING
PAGES TO SEE HOW IT GOT
THIS WAY.

THE TINY ARM OF *T. REX* WAS WASHED
AWAY FROM THE REST OF THE SKELETON AS
IT LAY IN A GENTLE STREAM.

PARTS OF THE SKULL BEGAN TO BREAK
AWAY AND WASH DOWN TOWARD THE HIPS.

shoulder bones that Kathy had dug away at.

The crew exposed the tops of the bones and began digging around the outer boundaries of the fossils, searching for bones and fragments that might have been scattered. Pretty soon, in a matter of a few days, we had a pretty clear idea of the layout of the skeleton, and it looked amazingly complete.

The body ended up on its left side, the neck arched back in death rigor. That's how most animal skeletons look, because the muscles of the neck and tail contract, curling up the body before those muscles and tissues decay. Even curled up, and without much of its tail, this was a huge skeleton, the biggest one I'd ever excavated. Looking at this animal gave us all chills. We were seeing more of *T. rex* than any living thing had seen in 65 million years. The skeleton was spread across an area of almost four hundred square feet and would have stretched more than thirty feet if the parts were joined together as in life.

To figure out all we can about how and where this animal died we'd need to sift through the debris for all kinds of fossils and study the lay of each bone from maps after the animal was completely excavated, something we haven't done yet.

And for this *T. rex*, we'll never know how it died. But right away we could tell something about where it was

THE HEAD SEPARATED
COMPLETELY FROM THE NECK

... AND FLOATED BACK
TOWARD THE HIPS.

buried and what happened to it. It either died in a stream or was washed into a stream after it died. It wasn't a gentle stream, because the animal's right leg was dislocated and moved back behind the skeleton. It might have taken a relatively good current to flip that big right leg over, though if the carcass had been bloated with gas as it rotted, the bones might have been more easily turned. The current quickly brought in sediment to cover the bone and keep it from eroding while it slowly fossilized.

The current in the sandy bottom of the stream channel wasn't so strong that it disturbed the largest elements of the skeleton. But it did wash many bones out of position. The left upper arm was pushed several feet away. That's the bone that Kathy Wankel found first. The four-and-one-half-foot-long skull had separated from the jaws, which still held several thick serrated teeth, each half a foot long.

The skull had rolled up against the pelvis. The right rear leg had toppled over, and its bones had separated. We found the left leg a year later when we began cleaning the pelvis. The pelvis had washed over it.

And we may still find the right arm and shoulder under the pelvis as we work on the back. Pat Leiggi did find a piece of the scapula bone of the shoulder in that block. Many of the small bones of the feet and the tip of

THE RIGHT LEG FLIPPED OVER
AROUND THE HEAD...

...AND THE ENDS OF THE TAIL
AND LEG FLOATED AWAY.

the tail seem to have washed so far downstream that we won't find them, though we do have the left hind foot, and one toe and two foot bones of the right foot.

Maybe if we followed the trail of the stream bed through the rock we'd find champsosaur vertebrae or wood fragments that would tell us a bit about which particular animals and plants lived, died, and were fossilized in the neighborhood of this *T. rex*. But for the moment we were set on finding all the bones we could of *T. rex* and getting those safely wrapped and back to the lab. Once we'd dug several feet out in every direction from the skeleton and found no more remains, we were satisfied that we'd come upon all the skeletal parts we'd be likely to find.

Though rock still surrounded the bones, our field-work was nearly done. All that remained was the hardest part, finding a way to get the bones we'd found packed and shipped back to the lab, since we can't do much fine preparation in the field.

Since a whole *T. rex* surrounded by rock is too big and

TOP LEFT: WE DUG AROUND AND DEEP UNDERNEATH THE WANKEL *T. REX* BEFORE JACKETING THE UNDERSIDE. HERE PAT IS USING A DRILL TO DIG UNDER A PARTIALLY JACKETED BUNDLE OF HIP BONES. WE CALL THIS TUNNELING PROCESS "PEDESTALING."

TOP RIGHT: THE FRONT-END LOADER HAS RAISED THE JACKET SO BOARDS CAN BE PLACED UNDERNEATH AND BOB HARMON AND BEA TAYLOR CAN APPLY PLASTER BANDAGES TO THE UNDERSIDE OF THE JACKET.

LEFT: I WORKED CAREFULLY ON SEPARATING THE SKULL OF THE WANKEL *T. REX* FROM THE PELVIS IT HAD ROLLED UP AGAINST SO WE COULD MAKE SEPARATE PLASTER JACKETS FOR EACH BONE GROUP.

heavy to move around the corner, let alone a couple of hundred miles, it had to be taken apart. Our challenge was to try to separate the skeleton into several discrete bundles without breaking apart bones.

Some bones, like those of the leg, were free and clear of other parts and could be wrapped into small individual packages. But for others, Pat had to make difficult decisions about where to separate the bones, especially where the vertebrae of *T. rex*'s back met its pelvis. Somehow Pat found a place to cut away the rock and divide those elements without damaging any bone.

Each group of bones had to be surrounded with tunnels that would allow us to get ropes and steel bands underneath to plaster the sides and bottoms of the bundles as well as the tops. As I mentioned before, this cutting around and below the fossils is called "pedestaling." It's hard, dirty work. You end up on your back, or your side, reaching deep under huge bones to scrape away at rock. There's not much room to maneuver, particularly with film crew looking over your shoulder

WE NEEDED A FRONT-END
LOADER TO GET THE BIGGER
BLOCKS OF BONES OUT OF
THE GROUND. THIS
MACHINE IS A TEREX,
PRONOUNCED *T. REX* BY
OUR CREW.

(and sometimes stepping on the fossils, and turning Pat white). This is not my favorite activity.

The only digging I did was separating the skull from the pelvis. The skull is the most important part of the anatomy because of what it can tell us about the animal's evolution and behavior. And it's the most delicate. This skull was so close to the pelvis that moving either might damage both. So if anyone was going to screw up, I wanted to be the one.

I began scraping with an awl until I found a little pathway of sandstone I could scrape away between the skull and hips. It was a relief to all of us that the skull and pelvis weren't flush against each other and so bound for damage when we moved them.

With a chisel I made a bunch of tunnels to make sure I had a good separation between the skull and hip bones. I needed enough space to get plaster on both the skull and hips, and room to move the skull, so the pelvis could then be removed. I ended up making tunnels big enough for me to crawl right through them.

THE FRONT-LOADER
TRANSFERRED THE FOSSIL
BUNDLES TO A TILT-BED
TRUCK FOR THE BUMPY
TWO-MILE RIDE TO THE
GRAVEL ROAD.

Meanwhile, others were cutting burlap strips and soaking them in plaster. They slapped them over the bone with a layer of paper towel in between to help separate them later. Then, as with every bone bundle, we lifted the skull up, plastered the last remaining patch on the bottom and set it aside to dry.

Once we'd moved the skull, we tried to make the smallest pelvic block possible, but even so, the pelvis, with the left hind leg, weighed three tons. The neck vertebrae formed a huge block, too, at least a ton. Together all of us couldn't move these blocks by hand, even to lift them for plastering the undersides, and we had no desire to try.

Fortunately Ed Westemeir, the Army Corps of Engineers' maintenance foreman, brought two of his expert front-end-loader men, Walt Murch and Kermit Flom, to haul the *T. rex* out. They came out July 2, less than a month after Pat had started the dig. Despite hailstorms and media plagues, Pat had everything ready to go right on schedule. Pat usually looks pretty relaxed, but he wasn't calm that day, trying to get that *T. rex* safely out of the ground.

Shelley's brother, Bill McKamey of Willimac Trucking in Great Falls, Montana, had driven his flatbed semi more than 300 miles to take the *T. rex* another 370 miles back to our museum in Bozeman. But since the road was too narrow to bring Bill's truck to the site, the bones had to be loaded first onto one of the Corp's tiltbed trucks. Tom Wankel brought out his two-ton grain truck to help.

All these volunteer drivers knew a lot more about building roads and fixing dams and docks than they did about dinosaurs. "Tri-sorus-pots" is what Kermit called *Triceratops*. I like his name better.

And we knew nothing about moving a fossil this big.

Turned out the Army Corps drivers were as good as Pat's crew at handling a *T. rex*, though we did have some scary moments.

Walt maneuvered the loader right up a short, steep incline to the top of the quarry. I took it as a good sign that the loader said "Terex" on the side. Walt lowered the loading fork to hoist the heaviest blocks slightly so Pat and the crew could put a pallet made from two-by-fours underneath, to support the blocks while they plastered the underside of the bones.

After the plaster dried, the loader's big yellow nylon straps were strung around the large bone packages on the pallets and cabled to the loader's bucket. The largest of the bundles, containing *T. rex*'s pelvis, weighed sixty-five hundred pounds. The second, with the neck vertebrae, was nearly as hefty.

Though the loader was built to carry that much weight, it wasn't designed to hoist irregularly shaped bundles like our jacketed fossils. One jacket tipped, the platform creaked, and the loader's back wheels tipped off the ground as the bundle was lowered back in place. For a moment we all had waking nightmares of a smashed *T. rex*.

By backing the tiltbed truck behind the loader and refitting the straps on the bundle, we made the job a bit less frightening. Walt and Kermit slid a dozen bundles of various sizes onto the tiltbed. While we kept a careful watch, Kermit inched the truck out onto the two-lane gravel refuge road where Bill McKamey had parked his semi. Walt transferred the bundles with the front-end loader, then covered them with tarps.

The *T. rex* move, begun at noon at the quarry, was completed at sunset as a storm broke overhead. Pat could finally relax. We stood by the road and watched *T. rex* roar off again across Montana for the first time in 65 million years. In the background, thunder boomed and the sky flashed with lightning. *T. rex* was leaving its world, and coming into ours, but not without a big fanfare.

BILL McKAMEY WAS WAITING AT THE ROAD WITH A SEMI TRUCK TO BRING *T. REX* BACK TO THE MUSEUM OF THE ROCKIES ON A STORMY NIGHT.

CHAPTER 3

DISCOVERING *T. REX*

THE TOWN OF JORDAN in eastern Montana has been the headquarters of *T. rex* country for almost a hundred years. But I knew I wouldn't find anyone who had known the original *T. rex* discoverer, the great Barnum Brown of the American Museum of Natural History in New York City, since Brown had stopped digging here in 1909.

The folks around Jordan surprised me, though, especially an energetic lady named Pauline ("Polly") Wischmann, a historian from the nearby town of Circle. Pauline hadn't met Brown, but she told me she'd written to Brown, and she sent me a copy of his brief response, from Guatemala City, dated March 12, 1953:

> I started work in the badlands of the Missouri River in the Spring of 1902 with Jordan on Big Dry Creek as headquarters and I continued working the region each summer until 1909.... My first discovery of a dinosaur [there] was the type skeleton of *Tyrannosaurus rex* at the old Max Sieber buffalo cabin on Hell Creek 16 miles northwest of Jordan.... The second *Tyrannosaurus rex* skeleton now mounted in the American Museum I found on the John Willis ranch on the Big Dry which I think was also Garfield County....

The discovery of *T. rex* was a lot more complicated and lucky than Brown's note suggests. What Brown didn't have time or inclination to write, I've learned from looking through old scientific papers and from Ralph Molnar. Ralph is an American paleontologist, though he's been living and working in Australia for many years. Ralph has spent more time studying *T. rex* than anyone alive, and he has also researched and written the history of its discovery.

As Ralph learned, the discovery of *T. rex* began with a paperweight. The weight was a tubular piece of fossil bone that belonged to William T. Hornaday, the director of the New York Zoological Society, at the turn of the twentieth century. Unlike modern-day zoo directors, Hornaday liked to hunt. He went to eastern Montana to shoot the last woodland buffalo in Garfield County (in 1886) for the Smithsonian's collections (Hornaday's buffaloes are now in our museum in Bozeman). And he had collected his fossil paperweight while hunting deer in Garfield County in 1901, without the slightest idea of what animal it belonged to.

Hornaday worked closely with the scientists of the American Museum, comparing fossil animals to exotic living creatures. So he showed his paperweight to young Barnum Brown, the museum's chief fossil collector. Brown was named for circus showman P.T. Barnum and had a flair for the theatrical himself. He grew up in Kansas but dressed like an eastern dude, sporting a raccoon coat, shiny boots, *pince-nez* glasses, and a gold-headed cane. He'd started digging fossils in college, at the University of Kansas, and even before graduation went to work for the American Museum. He stayed with the museum for sixty-six years, until his death in 1963 at the age of eighty-nine.

In that time he became fabled as a dinosaur hunter with an uncanny knack for finding fossils—it was said he could smell them. He found them throughout the West, and in Canada and India as well. Brown may have found more dinosaurs than anyone before or since.

With his eye, or nose, for potential headlines, as well as dinosaur fossils, Brown recognized Hornaday's souvenir as the core of a *Triceratops* horn. Brown also looked at photos of the territory of the fossil. To him they resembled the Wyoming badlands where he'd prospected for some of the last dinosaurs, 65 million years old.

The paperweight and photos are what spurred Brown

TOP: *T. REX* DISCOVERIES DON'T HAPPEN EVERY DAY AROUND JORDAN, MONTANA AS THIS MARQUEE ON A CLOSED JORDAN MOVIE THEATRE SUGGESTS, BUT THEY DO HAPPEN MORE FREQUENTLY THERE THAN ANYWHERE ELSE ON EARTH.

BOTTOM: BARNUM BROWN (LEFT) AND HENRY FAIRFIELD OSBORN (RIGHT) AS YOUNG *DIPLODOCUS* PROSPECTORS.

BARNUM BROWN WAS ONE
WELL-DRESSED FIELD MAN AS HE
IS HERE, PROSPECTING IN
SWEETWATER, MONTANA.

to go dinosaur digging along the Hell Creek Formation in 1902. It was a gamble on Brown's part, because even though dinosaur bones can be found everywhere in that country, geological explorers of the preceding fifty years had somehow missed them all.

Brown passed through Jordan. Today, Jordan isn't a big town, even by Montana standards. There's one big intersection in the middle of town; a few stores, bars, and motels; and a movie theatre that's been closed a few years (though it still advertises a documentary about local *T. rex* discoveries). A couple of hundred people, mostly ranchers, live nearby, friendly and proud to be from the capital of *T. rex* country.

But when Brown came in 1902, what he saw of Jordan was "three log houses nestling among the cottonwood trees." He collected plenty of fossils in the area, including a skeleton of a duckbilled dinosaur (*Anatotitan copei*) west of Jordan and a *Triceratops* skull near the town. And when he first arrived at the

Sieber ranch, where Hornaday had found the *Triceratops* horn, Brown came upon *T. rex* almost immediately, "before the cook's call for dinner."

In what Brown described as "flinty blue sandstone nearly as hard as granite" on the side of a hill, he came upon a small bit of bone the color of milky coffee. Digging into the hillside, his crew found more bones of the animal embedded in the sandstone. Standing out in bright contrast to the blue rock were parts of a hip girdle and hind limbs and a few backbones of a huge animal lying on its side. Where exactly the site was is hard to tell as Brown didn't like rival collectors to know where he was finding things. Since he was collecting for display, not for science, he didn't always keep very careful field notes. And he wasn't much for digging. He didn't like to get his fancy boots dirty.

Brown left the site, but over the next winter he began to sense the significance of his find. His crew didn't return until 1905. They dynamited great stretches of the cliff surface to expose more of the bone layer. It took many weeks before the continuation of the huge egg-shaped sandstone deposit that held the *T. rex* was found. Brown's workers dug a quarry thirty feet by twenty-one feet and twenty feet deep to extricate the huge skeleton. The largest of several rock blocks containing the animal weighed forty-one hundred pounds in its crate. Four horses were needed to pull it to the train in Miles City, Montana, more than a hundred miles away. The backbone, pelvis, hind limbs, and most of the tail were found, but the skeleton was far from complete, especially in the skull.

Brown thought he had dug up something new. But since big predatory dinosaurs were all but unknown then, he wasn't sure what he'd found. For instance, he would have been so surprised by *T. rex*'s small forelimbs that he wouldn't have recognized them as arms, had he come across them. Instead he thought a larger upper arm bone of a plant-eating dinosaur found nearby belonged to *T. rex*.

What's more, Brown didn't realize that the creature

we call *T. rex* had already been discovered, twice, one of those times by himself. In 1900, north of the Cheyenne River in western Wyoming, Brown had found the lower jaw and backbones of what he thought was a large meat-eater.

Identifying the Wyoming and Montana carnivores was an honor that went to Brown's boss, Henry Fairfield Osborn. Osborn was then the director of the vertebrate paleontology department at the American Museum, but he would soon (in 1908) become its most famous president.

Osborn was an upper-crust Princeton man. He got hooked on collecting fossils when he found dinosaurs in Wyoming on a geology course field trip. (He also got his good-sized nose so sunburned that a prospector he encountered told him, "Young feller—either you had better pull out the brim of your hat or pull in your nose.") Osborn became one of the leading scientists of his day, and apparently he liked to let people know that. He had plenty to boast about. Osborn's name appears on an incredible 940 scientific and popular articles and books.

Even before Brown finished excavating his giant

predator find in Montana in 1905, Osborn had named it *Tyrannosaurus rex*. "I propose to make this animal the type of the new genus, *Tyrannosaurus*, in reference to its size, which greatly exceeds that of any carnivorous land animal hitherto described." Osborn added the species name *rex*, with the same motivation, creating the "tyrant lizard king." The king measured thirty-nine feet long and stood nineteen feet high by Osborn's calculations. Osborn could not help celebrating this spectacular possession of his museum: "This animal is in fact the ne plus ultra of the evolution of the large carnivorous dinosaurs: in brief it is entitled to the royal and high sounding group name which I have applied to it."

Lots written about dinosaurs then sounds silly to us today. But the turn of the century was a time when people didn't know the age of the earth and what animals lived on it when.

Reporters wrote about the *T. rex* "who munched giant amphibians and elephant à la naturel." Osborn himself wrote of *T. rex* being "3 or 4 million years" old. Yet Osborn did the best he could with the information then available. He and other scientists of the time already understood, as we do, that *T. rex* lived at the end of the last dinosaur era, that it did not live at any time

near the days of *Stegosaurus* and "*Brontosaurus*," and that it was closely related to both living reptiles and birds.

The pieces that Brown had found in Wyoming in 1900 (now at the Museum of Natural History in London), of a meat eater very much like *T. rex* but a bit less robust, Osborn named *Dynamosaurus imperiosus* ("imperial powerful lizard"). In 1906, Osborn changed his mind and identified that animal as *T. rex* also, and the *Dynamosaurus* name was pretty much abandoned.

Brown kept looking for *T. rex* in the Hell Creek Formation, and in 1905 he found the right and left hind limbs of a *T. rex* smaller than his first Montana discovery.

Two years later Brown headed to the ranch of John Willis, a hunting friend of Teddy Roosevelt's, and badlands "unsurveyed and unmapped," according to Osborn. Brown was looking for another *T. rex*, but at first he had no luck. The way he told the story (to his longtime assistant, Roland T. Bird), "I searched the badlands for a month without finding a thing. I had given up: then I decided to give it one more day. Turned out to be my day: he's still the biggest *Tyrannosaurus* ever found." For more than half a century, Brown was right.

On the fateful day in 1907, near the summit of a hill, Brown came upon an exposed bit of what proved to be the huge skull of a *T. rex*, almost four feet long with eight-inch teeth in its massive jaws, lying on its side. "The skull alone is worth a summer's field work for it is perfect," Barnum Brown wrote on August 10, 1908. Brown had more, almost a complete animal, minus the fore and hind limbs and bits of the tail. The skeleton took nearly a summer's work to excavate and was removed by a team of horses in blocks of stone that weighed fifteen tons. The lower jaws were in a block weighing five tons, according to Brown.

In his wife's book *I Married a Dinosaur*, Brown recalled this *T. rex* as "my favorite child," and that it took him "two summers to dig the fellow up and transport his remains by wagon one hundred and twenty miles to the nearest railroad." "Even after we got it back to the

THE ELEVEN AND ONLY *T. REXES*

1. The first discovery of a *T. rex*, in 1900 by Barnum Brown in western Wyoming. Originally named *Dynamosaurus imperiosus* by Dr. Henry Fairfield Osborn, the specimen is now in the British Museum of Natural History in London.

2. The type specimen of *T. rex*, found in 1902 by Barnum Brown in Garfield County, Montana, named by Dr. Henry Fairfield Osborn in 1905. On display at the Carnegie Museum in Pittsburgh, Pennsylvania, this *T. rex* is 50 percent complete.

3. *T. rex* excavated in 1907 and 1908 by Barnum Brown in Garfield County, Montana. Missing fore and hind limbs. On display at the American Museum of Natural History in New York.

4. *T. rex* found in 1966 by Harley Garbani in Garfield County, Montana, for the Los Angeles County Museum of Natural History, where its skull is on display. The specimen is 60 percent complete. Garbani also found parts of two other *T. rexes* in Garfield County.

5. *T. rex* excavated in 1980 by scientists from the South Dakota School of Mines in northwestern South Dakota. Found by rancher Jennings Flowden, the skeleton is 40 percent complete. The skull is on display in the university's museum.

6. In 1981, scientists at the Royal Tyrrell Museum of Palaeontology in Drumheller, Alberta, collected 30 percent of a *T. rex*, originally located by Charles M. Sternberg in central Alberta in 1946.

7. "Black Beauty," a second, more complete *T. rex* from southern Alberta, was found in 1981 by high school students and was mounted at the Royal Tyrrell Museum in 1992.

8. In 1981, Mick Hager of the Museum of the Rockies collected 40 percent—the whole hind end, but not the rib cage, skull, or neck—of a *T. rex* in eastern Montana.

9. In 1987, *T. rex* remains were discovered by collector Stan Sacrison outside Buffalo, South Dakota. In April 1992, Pete Larson of the Black Hills Institute in Hill City, South Dakota, excavated the site and found a *T. rex* at least 60 percent complete, including the first known specimens of the animal's tip-of-tail vertebrae.

10. In September 1988, Kathy Wankel discovered a *T. rex*, nearly 90 percent complete, in McCone County, eastern Montana. It was excavated in June 1990 by a Museum of the Rockies crew. The specimen is on display during preparation at the Museum of the Rockies in Bozeman, Montana.

11. On August 12, 1990, Susan Hendrickson of the Black Hills Institute discovered a *T. rex* specimen in western South Dakota. Excavated that month, the fossil is now impounded by the FBI at the South Dakota School of Mines. "Sue" is at least 90 percent complete and is the largest known *T. rex*. If the institute wins custody of the disputed *T. rex*, it will be displayed in the institute's planned museum, as will *T. rex* discovery #9.

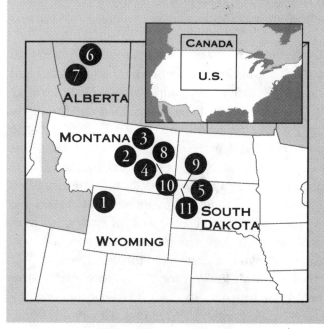

museum our troubles were far from over," Brown told his assistant, Roland Bird. "The skull and limbs worked out easily but the pelvis and some of the connecting vertebrae were encased in a hard iron concretion," ironstone that Brown's assistant had to "take out in the yard and knock off with hammer and chisel." It took three summers for Brown and his crew to dig out all of the giant, working with picks and dynamite.

Osborn wrote five papers on *T. rex*, based on Brown's finds. When these finds were combined, they left only the hands and tail of *T. rex* a mystery. Osborn guessed about those, and guessed wrong. He gave *T. rex* three-fingered hands, like those of an *Allosaurus*, a huge predator nearly 80 million years more ancient than *T. rex*. And Osborn figured *T. rex*'s tail several feet too long. But without more *T. rex* to go by, Osborn had made some pretty reasonable speculations.

The second successful *T. rex* hunter wasn't a paleontologist, and he didn't work for an eastern museum. Harley Garbani is a plumber. He's in his seventies now and lives in a trailer park in the San Jacinto Valley, south of Los Angeles. I've met him a few times out in Montana, and it was a great honor for me. That's because Harley is also the world's leading discoverer of *Tyrannosaurus rex*.

Harley grew up on a farm in the San Jacinto Valley. Like me, he found his first fossil when he was just a kid, eight or nine. His wasn't a dinosaur—it was a piece of camel leg. That find hooked Harley on fossil collecting. He kept hunting through high school and after marriage and World War II, in the deserts of southern California and northern Mexico. He got so good at finding stuff, the Los Angeles County Museum of Natural History hired him to collect for them in the summertime. In 1965, the museum had the good sense to send Harley, his wife, and daughter to look around eastern Montana where Barnum Brown had discovered the first *T. rex*.

Harley wasn't looking for *T. rex* only. "I'd collect every fossil from mice to *T. rex*. I loved just beatin' the bush." Beating around Lester Engdhal's ranch in the

THE NEXT GREAT *T. REX* HUNTER AFTER BROWN IS THIS GENTLEMAN, CALIFORNIA PLUMBER HARLEY GARBANI.

Hell Creek Formation badlands north of Jordan on July 27, 1966, Harley came upon "the damn thing"—some large foot bones sticking out of a bank. Right away, Harley knew he was onto *T. rex*. (And more, he also found parts of a juvenile *T. rex* buried with it!)

"I was pretty excited. I didn't figure another of those suckers would be found," Harley said. And he was pretty sure he'd found one. "I'd seen enough of *Trachodon* [a dubiously named duckbill] to know this was something bigger."

"Possible very adult *Tyrannosarus* [sic] limb, tarsal & toes," Harley wrote in his field notes that day. By the summer of 1970, Harley's crew of local volunteers and high-schoolers had dug up a well-preserved *T. rex*, but in a jumble of bones that made up only about 30 percent of the animal's skeleton. Those remains did include the outstanding skull, nearly 75 percent complete.

The skull Harley excavated is now on display at the Los Angeles County Museum of Natural History, and it is the biggest *T. rex* head on view anywhere, nearly five feet long.

Harley found still more *T. rex* when he went back to the Hell Creek Formation in 1977 prospecting for the University of California, Berkeley. Berkeley's expert paleontologist Bill Clemens (Bill works primarily on fossil mammals) had told him of a promising site on the slope of a hill. Harley investigated and found the upper jawbone (maxilla) of a *T. rex*, even bigger than the first he'd found a decade earlier. He found one huge piece of a leg and a lot of fragments, but it didn't add up to much of a dinosaur.

In 1982, Harley came upon yet another *T. rex* in a hillside in the Hell Creek Formation while he was working for the University of California. This one had two "good-sized" upper and lower jaws. As for why he's found three *T. rex*es when less than a dozen are known in a century of prospecting, Harley says, "I'm just a pretty lucky fella. They're so rare, it's just a privilege to find them."

Phil Bjork of the South Dakota School of Mines found

a *T. rex* in northwestern South Dakota in 1980. That one
is about 40 percent complete, with most of the skeleton
coming from the skull and neck. The texture of the
bone is weak and in a poor state of preservation.

In eastern Montana the following year, Mick Hager,
former director of the Museum of the Rockies, found
parts of a *T. rex*—hind legs and most of a tail. Those
bones are in our study collection at the museum. We
also have an isolated skull, disarticulated, which was
collected by an archaeologist digging in the Hell Creek
Formation in the 1960s.

*T. rex*es have been found north of Montana in Alberta,
Canada. The first was actually discovered in central

Alberta in 1946 by Charles M. Sternberg. Sternberg and his father and two brothers were the greatest fossil-collecting family of all time. The father, Charles H. Sternberg, was a minister's son from rural New York who wrote colorfully in diaries of time-traveling mentally into dinosaur days. It sounds wacky, but Sternberg senior turned into Barnum Brown's chief competitor as a dinosaur finder.

The *T. rex* bones C. M. Sternberg found were stuck in hard ironstone near the top of a bluff outside the town of Huxley. Sternberg couldn't get at them, and no one else was able to for decades.

In 1960 another great paleontologist, Wann Langston, came to the Huxley *T. rex* site. Wann was prospecting for the National Museum at Ottawa, Canada, when he saw parts of a badly broken-up *T. rex* skull at the base of the same cliff where Sternberg found the *T. rex* skeleton. Chances are both finds come from the same animal. So the rest of the *T. rex* skull may still be in the cliff top.

In 1981, a crew from the world's largest fossil museum, the Royal Tyrrell Museum of Palaeontology in Drumheller, Alberta, came to excavate the Huxley *T. rex*. With jackhammers and pneumatic drills they pulled out the pelvis, hind limbs, and backbone from neck to tail of a *T. rex*. This was a big *T. rex*, probably forty feet long. The Tyrrell crew tried to get at the skull end of the neck in the back wall of the quarry. But to do so safely they would have had to move another ninety feet of rock overburden, an overhwelming job for a few folks with hammers and drills.

My friend Phil Currie thinks he might be able to move that rock to get at the *T. rex* skull, and I don't doubt him. Phil is the Tyrrell's dinosaur paleontologist. He's one of the world's experts on predatory dinosaurs and an expert digger. Lately, he's gone to Australia to learn dynamiting techniques from Tom Rich, a dinosaur paleontologist at the Victoria Museum in Melbourne who blasts holes in cliff faces in southern Australia to find dinosaur remains. Until Phil dynamites and mines,

PHIL CURRIE IS THE TYRRELL'S SENIOR DINOSAUR SCIENTIST, ONE OF THE WORLD'S EXPERTS ON CARNIVOROUS DINOSAURS, INCLUDING *T. REX*, AND MY PAL.

no one's sure just how much of *T. rex*'s skull lies in the Huxley cliff.

One other Canadian *T. rex*, "Black Beauty," was found in 1981 in southern Alberta by three high school kids. Once you get out to these foothills near the Crow's Nest River, these *T. rex* bones would be pretty hard to miss. Manganese has turned them deep black in the middle of reddish-white sandstone. Like the other Canadian *T. rex*, this one was also locked deep in horribly hard rock. But this *T. rex* was close enough to ground level that it could be taken out by bulldozer. The skeleton is now on display at the Royal Tyrrell Museum. At first no one thought there was much of a skull left, but when one fossil block was flipped over and cleaned in the lab, it turned out to hold an almost intact skull. The size of the beautiful black head, nearly five feet long, suggests it belonged to a forty-foot-long *T. rex*.

The biggest *T. rex* of them all is the newest find, "Sue." Sue is named for Susan Hendrickson, a former field collector for a company that finds, cleans, and sells fossils—the Black Hills Institute of Geological Research in Hill City, South Dakota.

In August of 1990, a month after we took Kathy Wankel's *T. rex* out of the ground in Montana, Susan was out in the South Dakota badlands, north of the town of Faith, digging at a duckbill quarry with a Black Hills crew. Work had stopped for a day while their only vehicle was taken to town for repairs. Susan went for a walk to look at a hillside near where the crew had found some pieces of a *Triceratops* skull a few miles from camp. She saw some vertebrae, still articulated, some ribs, and another large bone weathering out of the side of a cliff. Susan took two huge vertebrae, several inches wide, and brought them back to the crew. They walked back to the hillside where she found them and saw part of the pelvis and several backbones sticking out of the ground.

From the size and shape of the bones, Pete Larson, president of Black Hills Institute, knew at once Susan had found a *T. rex*. Three Black Hills workers went to

MEET "SUE" THE *T. REX*, OR AT LEAST HER HEAD, AS IT WAS EXPOSED IN SOUTH DAKOTA.

work on digging and made a pretty big hole, twenty-nine feet deep and twenty-five feet wide. In two weeks they had the whole *T. rex* out in blocks that weighed as much as nine thousand pounds.

Back at their lab, they realized they had most of a *T. rex*, including only the second front limb ever found (ours was the first) and thirty-five of the tail vertebrae. That's more than twice what we've got of our *T. rex* tail, and more tailbones than anyone has found (though not as many as Osborn invented for the American Museum *T. rex* display). Although our *T. rex* has more arm bones than Sue or any other *T. rex*, Sue has many parts our *T. rex* lacks—a complete foot, many more ribs, more of the skull, and all that tail.

With nearly all of Sue's tail intact, for the first time we can estimate accurately how long a *T. rex* was. The estimates before ranged from thirty-four feet to more than fifty feet. Now we know forty feet was more like it. And that's the length of the biggest *T. rex* in the world, so far. Sue is at least 5 percent bigger than any other *T. rex* known. Sue's thigh bone (femur) measures fifty-four inches long. That's three inches longer than those on the second and bigger of the Montana *T. rex*es Barnum Brown found, the *T. rex* in the American Museum in New York (according to Dr. Osborn's generous measurements).

Sue's skull is in terrific shape, black teeth still in massive jaws that aren't distorted. And her bones show all sorts of injuries that had healed over in the course of what must have been a long, rough life. Extra bone growth along the lower left shin indicates to Pete Larson that Sue had a broken leg that healed over, though it might also have suffered from osteoporosis, arthritis, or other bone diseases that we know dinosaurs had. There is a hole that might have been a drain for a cheek infection. There are parallel grooves along the pelvis, maybe from the raking bite or slash of another *T. rex*. Two vertebrae near the end of the tail were probably broken. They were found fused with extra bone around them. There is a rib incompletely healed from a fracture. And there is a tooth still embedded in a rib, from a bite by another *T. rex* (which may have come after Sue's death). As I'm editing in 1994, Sue's cleaning is far from finished, and the specimen has been impounded by the FBI in an ownership dispute (more on that later), but it is clear that Sue is more than 90 percent complete.

A lot more came out of that quarry than just Sue. The *T. rex*'s body seems to have acted as a trap for lots of leaves and the bones of other dinosaurs, from a little hypsilophodont browser to parts of three other *T. rex*es.

Vertebrae and other bones of a big duckbill, an *Edmontosaurus*, were found with Sue. The bones looked etched and partially coated with ironstone. Pete Larson thinks they were stomach contents of a *T. rex*—remains of a partially digested meal. He also thinks he's found what's left from fully digested dinners—dinosaur excrement, which he's having analyzed.

Those conclusions may be difficult to substantiate. But what's more certain, and exciting, are the bones of other *T. rex*es found along with Sue's. There are the ends of two right leg bones of a second *T. rex*, about 70 percent as long as Sue. This could be part of a subadult, a teenaged *T. rex*. Something, maybe another *T. rex*, gnawed on this one, as it shows bite marks on its tibia and fibula.

The left lachrymal, a bone at the front of the eye, of a third *T. rex* was found, one that would fit a skull about a foot and a half long, less than a third the size of Sue's—perhaps from a *T. rex* youngster.

Another skull bone was found that came from a skull less than a foot long—a *T. rex* baby! Of course, identifications from a bone or two are risky, and skull proportions no doubt changed as these animals matured. But Tyrrell expert Phil Currie, who knows *T. rex* skulls as well as anyone, finds these measurements believable.

Was this the burial ground for a family of *T. rex*es—parent and three children of different ages? It's possible. But it's more likely that the animals died at different times and places and washed into the same stream bone-trap.

bone-trap.

Sue's excavation also produced some bones from other predators we don't see much of in the Hell Creek rocks. The dig turned up bits of smaller dinosaur carnivores, possibly a dromeosaur and a caenagnathid, as well as the skull of a turtle.

And in the spring of 1992, Pete Larson made yet another *T. rex* find. Actually, Stan Sacrison did. Stan's a fossil collector from Buffalo, South Dakota. Stan had found some eight feet of linked vertebrae leading into a hillside outside town five years before. Pete had gone out with Stan in January of 1992 to look at the site. As he told me, "I walked down the hill and right away said, 'Holy cow! Another *T. rex*!'" For Pete the give-aways were "the distinctive vertebrae. They were hollow with a concave surface. They had to be from a big predator and they were too big for *Albertosaurus*."

Pete came back as early in spring as he could to dig up the site with his crew. They came upon the pelvis and some backbones that led to tailbones underground. The bones were locked in a very hard matrix of ironstone, but Pete managed to remove several blocks. His preparators are busy cleaning the stuff in the lab, but already they've found twenty-six loose teeth with roots, perfect for my former student Greg Erickson to work on for his bone growth research.

Pete also found a complete leg, at least ten ribs, a lot of the pelvis, and most of a skull, disarticulated. Now he's got one *T. rex* skull in each state of preservation— an intact skull from Sue and one in pieces, like ours. That's ideal for measuring individual bones and seeing how they fit together.

But this *T. rex*'s skull wouldn't look quite like Sue's if it were put back together. The whole animal is quite a bit more lightly built than Sue, more along the lines of our own *T. rex*. The difference in build may have implications for judging dinosaur genders, and I'll get to that in Chapter 8. For now, let's just say if the first Black Hills *T. rex* was "Sue," Pete thinks this one is "Stan."

It seems like *T. rex*es are popping up everywhere. In May of 1992, Charles Fricke was out walking his dog in an empty lot next to his house in a subdivision outside Denver when he found a big bone. Mr. Fricke was pretty excited (and so probably was his dog). He's an amateur fossil collector, and he knew he'd found a fossil. He

PETE LARSON AND CREW TAKE A PHOTO OPPORTUNITY AROUND THE SKULL OF SUE.

called the Denver Museum of Natural History, which sent a team out to collect the fossil. At the museum, preparator Ken Carpenter identified it as the distal end of the tibia (that's the bottom of the shin bone) of a *T. rex*. Museum crews dug farther in the lot, and they've found a femur (thigh bone), shoulder and ankle bones, stomach ribs, and parts of hip and leg bones. The find further extends the known range of *T. rex* in the American West.

I'm not surprised to hear there were *T. rex*es in Colorado, only that it's taken people this long to find parts of two of them. Now that there are a lot more knowledgeable people like Charles Fricke out looking for fossils, we're bound to find lots more *T. rex*es wherever in the West rock the age of the Hell Creek Formation is exposed.

I've heard of three more *T. rex* discoveries in the Dakotas over the summer of 1993, one of them found by Stan Sacrison. At least one of these finds may be a good skeleton.

What concerns me is that not all fossil collectors are as public spirited as Charles Fricke or Kathy Wankel. These days, dinosaurs are pretty fashionable, and a lot of people are collecting fossils to sell them. Some of these commercial fossil collectors cause serious problems for paleontologists like me. They aren't very careful about excavating valuable fossils and keeping notes. They cater to the big market for fossils as decorator items and get high prices from individual collectors and foreign museums. Scientists like me and museums like the Museum of the Rockies can't pay those prices. And private landholders don't want to let scientists go after their fossils when a commercial collector will pay them a lot more to make them into trophies.

But there are others like Canada Fossil, a commercial collecting outfit that has been very cooperative with scientists. They sell only to museums, not to individuals. They never sell the "type," or namesake specimen of an animal, but give those to scientists. And they take great notes. They let us see all those notes and their bone discoveries, and they make casts of their finds if we need them even before they go to museums.

Pete Larson of the Black Hills Institute knows first-hand the mess you can get into collecting fossils. Pete was schooled in paleontology (he stopped just short of a master's degree). Like many commercial collectors, however, he's upset some paleontologists and government officials. They question whether he's obtained all his specimens from land where he had permission to dig.

Those doubts about Pete's collecting led to a bizarre incident on May 14, 1992, when thirty FBI agents turned up at Black Hills Institute and confiscated Sue, the world's biggest *T. rex*.

Some folks have said Sue was found inside the Cheyenne River Sioux Reservation and belongs to the Sioux, though Pete says he paid $5,000 for the rights to excavate the site from the landowner, a member of the tribe. "Will the Sioux sue for Sue?" wrote a clever *Newsweek* caption writer.

They didn't. But the tribe's leader did ask the assistant U.S. attorney in Pierre, South Dakota, to investigate Pete's collection of Sue. The U.S. attorney said Pete's taking Sue was a violation of the Federal Antiquities Act, which "prohibits outsiders from making contracts with

people of the Indian community without Federal permission."

The U.S. attorney thinks the land is the government's because the landholder had placed it in trust with the U.S. government. According to Pete, this twenty-year trust, which expired in 1992, is a tax advantage, offered Native Americans for their lands, that should not affect fossil collection. Though the landowner still had mineral rights, if he sold an antiquity he was supposed to have federal permission. Pete should have gotten an antiquities permit before he started digging.

Whether *T. rex* is an antiquity is another story. It's certainly old, but the law was designed to protect archaeological artifacts. I wouldn't mind a bit if there was a clear new law protecting fossils found on state and federal lands. Some paleontologists are calling for a law nationalizing all fossils, on private or public lands, such as now exists in the province of Alberta in Canada, one of the richest places for dinosaur fossils in the world.

As I'm editing the paperback edition in early 1994, Pete's just been indicted by a grand jury on charges including theft of government property. Sue the *T. rex* is under lock and key in a warehouse at the South Dakota School of Mines. It's going to take a judge to spring her.

Pete's the one who has sued, both the government and the Sioux, to get Sue back. The Sioux asked me long before the FBI stepped in to look at the land where Sue was found. I told them I thought it was on Sioux land. Other than that, I've not had a lot to do with the mess.

If you ask me, and nobody has, the whole deal is pretty ridiculous. Fossils should be public property, and entrusted to scientific institutions to study. Period. And until some judge settles the question of who gets Sue, it could have stayed just where it was, in Pete's lab. Nobody's going to walk off with a tyrannosaur.

Whoever gets her, I'd be relieved to know that the bones of Sue, like those of other *T. rex*es, will be around for us to study.

CHAPTER 4

THE IMAGE OF *T. REX*: THE MAKING OF A MONSTER

THE MAN WHO NAMED *T. rex*, Henry Fairfield Osborn, is also the one who gave several generations of scientists, artists, moviemakers, and so most of us, our sense of what *T. rex* looked like and acted like—ferocious, upright, and tail dragging. That was the pose of the best-known *T. rex* skeleton of them all, the one that has thrilled so many youngsters and inspired a bunch of them (including Harvard University's Stephen Jay Gould) to become paleontologists.

That's how *T. rex* stood for most of this century at the American Museum of Natural History in New York City. But it isn't how scientists now think of *T. rex*. And it wasn't the impression Osborn intended.

Osborn did envision *T. rex* as the top dog among dinosaur hunters, "the chief exterminator of *Triceratops*." Its massive teeth, "pointed like daggers," were used to "terrorize the other animals that lived at the same time." *T. rex* was vicious, in Osborn's mind, in keeping with its primitive nature, "the tendency of the older forms to be the more quarrelsome and wage their combats with greater persistence."

Osborn thought *T. rex* walked upright and slept lying down. Its big T-shaped pubic bone, Osborn thought, would have been the anchor for heavy abdominal muscles. Those gut muscles held *T. rex* up when walking.

Though he thought *T. rex* walked upright, Osborn also believed *T. rex* was a fast runner. "For an animal of

this size, *Tyrannosaurus* was unquestionably fleet of foot." So also thought *T. rex*'s discoverer, Barnum Brown, who imagined *T. rex* as "active and swift of movement when the occasion arose."

Osborn thought its forelimbs, no longer than ours, were "apparently of very little use," except as "meat hooks." He did have one other interesting idea for how they might have been used—for stroking its partner during mating.

Henry Fairfield Osborn was not just a great and influential scientist, he was a smart administrator. As a scientist at the American Museum, and in the fifteen years he was its president (1908–1923), he made it into a world-class institution, in part by being a showman. He loved spectacular displays, and he had one in mind for *T. rex*. When he named *T. rex* in 1905, Osborn, then a senior researcher, promptly had what was available of the animal (hind limbs of yet another *T. rex* that Brown had found that year) mounted in a stunning but unlikely upright pose that made the legs stand fifteen feet from hip to floor. In December of 1906, even before the partial skull of *T. rex*'s namesake specimen was prepared, the legs of *T. rex* were put on display beside an *Apatosaurus* (then known as *Brontosaurus* and given a *Camarasaurus*-like head). The *New York Times* headlined its feature on the display of the mount "The Prize Fighter of Antiquity Discovered and Restored." Professor Osborn was credited with the discovery of the "swift two-footed tyrant."

With the two *T. rex* skeletons Brown had dug in the Hell Creek Formation, Osborn had enough fossil material to describe and recreate most of an entire *T. rex* skeleton. He immmediately set about doing so. The obstacles were great. Here were animals nearly twenty feet high and twice as long, whose heavy bones needed even heavier iron braces to support them. Once the metal frame was attached it was impossible to reposition the bones.

With *T. rex*, Osborn tried a method never before attempted for making dinosaur mounts. He had sculptor Erwin Christman make movable one-sixth-scale models of Brown's two Montana *T. rex*es and posed them in a variety of dynamic hunting tableaus. Osborn suggested three poses, showing *T. rex* lithe and agile enough to lift a leg like a chorus dancer. But when he saw the models of his high-stepping *T. rex*, Osborn wasn't satisfied that he'd come up with a design sufficiently stable and dramatic.

Osborn turned for advice, as he had in the past, to Raymond L. Ditmars, the curator of reptiles at the Bronx Zoo. Ditmars studied how large living lizards hunted and concluded *T. rex* would have hunted in the same manner—as Osborn said, with a "convulsive single spring and tooth grip that epitomizes the combat of

reptiles from that of all mammals."

With new ideas from Ditmars, Osborn decided to have two *T. rex* models arranged in battle over the carcass of a third, based on another less complete specimen Brown had discovered. Osborn had one model squatting over the carcass while the other reared up to menace its competitor. The squatting pose was selected not just to reflect Ditmars' impression of how a giant lizard might arrange itself while feeding; Osborn wanted the skull and pelvis, two of the most dramatic skeletal features of *T. rex*, close to the visitor's eye.

Unfortunately, the heavy iron rods and braces needed to support the full-sized *T. rex* made it incapable of such off-balanced posing. So Osborn reluctantly had to settle for the static upright pose of *T. rex* that for so long shaped our view of how the creature would have looked in life. Osborn wrote that the mount was "more effective with the feet closer together, the legs straighter and the body more erect." *T. rex* might have ended up even more upright, but the ceiling of the hall into which it was first placed required posing it six or seven feet

THIS IS HOW OSBORN WANTED *T. REX* TO LOOK. THESE POSEABLE MODELS WERE DESIGNED UNDER HIS SUPERVISION BUT PROVED TOO DIFFICULT AND EXPENSIVE TO BUILD.

short of the most upright posture possible.

To make *T. rex* look whole and stand steady, Osborn made other creative decisions. He added several feet of imaginary lizardlike tail, which also helped to stabilize the stand-up *T. rex*. For the missing forelimbs, he substituted those of the closest-known meat eater to *T. rex* in size, *Allosaurus*. That was a mistaken guess, as we later found out. In 1914, Canadian paleontologist Lawrence Lambe demonstrated that tyrannosaurs—in his discussion, *Albertosaurus*—probably had two fingers. As it happened, *Allosaurus* had three digits on each hand. *T. rex* had only two, as we proved for the first time with Kathy Wankel's *T. rex*.

Osborn's dream, a dynamic grouping of *T. rex* models, was supplanted by a full-sized and conservatively posed upright *T. rex* skeleton at the American Museum of Natural History, with the bones of another in storage. The museum's collection was narrowed to one in 1940 when World War II broke out in Europe. Concerned that German bombers might destroy the valuable skeletons, the American Museum sold the first Montana *T. rex*, the namesake of the species, to the Carnegie Museum in Pittsburgh, where it was mounted for the first time and where it still stands.

So, for generations, in its most famous and influential representation, *T. rex* has been viewed as a stiff, long-tailed, lumbering beast. That reputation was spread by the first and foremost of American paleontological painters, Charles R. Knight (1874–1953). Knight was a talented and careful artist who relied heavily on scientific opinion in his painting. He thought of the dinosaur not as a fossil, but as "an animated, breathing, moving machine," and studied living animals in zoos and in the wild to help him conceive how dinosaurs might have moved. Knight also talked to Osborn, Brown, and Matthew, then sketched the bones and mounted skeletons and made clay models before attempting any restoration scene.

Knight spent much of his adult life working for the American Museum of Natural History. Shortly after the

CIRCULATION
2,000,000

BIG NEWS

SECOND
EDITION

Published by Sinclair Refining Company (Inc.), 45 Nassau Street, New York, N. Y.

SINCLAIR

CUTIE CUTS CUTICLE. Manicuring has entered the field of big business at the World's Fair, judging from this photo of Julia Lyons, who is engaged in keeping Tyrannosaurus Rex, King of the Tyrant Reptiles, in shape for his daily battle with his ancient foe, Triceratops, at the Sinclair Dinosaur Exhibit. While these strange dinosaurs roamed the earth millions of years ago, Nature was mellowing and filtering the crudes from which are made today's Sinclair Opaline and Sinclair Pennsylvania Motor Oils.

legs-only *Tyrannosaurus* first went on exhibit there in 1906, Knight produced a painting of a *T. rex* facing down a family of *Triceratops*es. Osborn advised Knight closely on the work, so it shows many of Osborn's goofs—an upright pose, three-fingered hands, and a long tail, barely off the ground. The eye was placed too far forward in the head. But that's not because, as I once thought, Osborn and Knight put it into the wrong hole in the skull of *T. rex*, one in front of the actual eye socket.

Knight painted *T. rex* again, for *National Geographic* (in 1942) and in his masterpiece, a set of murals of life's history for the Field Museum of Natural History in Chicago (1926–1930). That *T. rex* mural still inspires visitors there. Thirty years ago paleontologist Phil Currie saw it when he was a young kid. He credits it as one of his first inspirations to study dinosaurs.

Knight's Chicago *T. rex* is seen in a profile view, leaning forward with tail raised, squaring off against a *Triceratops*. Today, we'd make the legs more birdlike and the snout longer, and we might make the body lean forward more, and raise the tail farther. But at least Knight got *T. rex* off his haunches. A second *T. rex* in the background is a lot less dynamic. It stands upright, its long, fat tail on the ground. Both animals' necks lack the S-shaped curves now known to be a feature of all dinosaurs' anatomy. And both *T. rex*es look a lot chunkier than we think they were.

Right or wrong, Knight's *T. rex*es were the ones that people took for real. They were reproduced in many magazines and books. Knight's *T. rex* was the one in the dime novel fantasies adapted from the work of Edgar Rice Burroughs, the man who wrote *Tarzan*.

And Knight's *T. rex* was brought to life, briefly, in 1933. It showed up as one of the dinosaurs in the Sinclair Refining Company's dinosaur exhibit at the Chicago World's Fair. Sixteen million people came to see the twenty-five-foot-long roaring, animated *T. rex* and five other giant dinosaurs. *T. rex* was the main bad guy, described as a "giant nut-cracker" in the Sinclair publicity. P. G. Alen, an Indiana model maker for Hollywood films, based the first robot *T. rex* on Knight's Field Museum mural and the World's Fair sculpture it inspired. Alen stuck to Knight's vision, down to the three-fingered hands.

Also in 1933, *T. rex* looked the liveliest it had in 65 million years, at least until the giant gorilla killed it in *King Kong*. That *T. rex* was a rubber cast of a model made by sculptor Marcel Delgado, straight from Knight's first painting.

T. rex was a lumbering beast in *King Kong*, and a loser. And so it has been in many films, from *The Valley of Gwangi* (1969—that dinosaur was actually a *T. rex–Allosaurus* hybrid) to *The Land Before Time* (1988). It has even been portrayed by a real-life lizard blown up, such as the iguana turned *T. rex* in *King Dinosaurs* (1955). In most of these films, the monster dinosaur is Knight's original *T. rex*, including the three-fingered hand. Writer Don Glut, an expert on dinosaurs' influence on popular culture (and my source for much of this information on *T. rex*'s image), quotes Walt Disney as telling *T. rex* discoverer Barnum Brown he preferred an extra finger on *T. rex* because it "looked better that way."

From the stout, upright *T. rex* in Rudoph Zallinger's famous mural, *The Age of Reptiles* (1953), at Yale University's Peabody Museum to bulky sculptural restorations that stood at the New York World's Fair–and still stand in museums from Boston to Queensland, Australia—the old, inaccurate images of *T. rex* have persisted. Since they are effective and well-made images, and museums are slow to update exhibits, these *T. rexes* persist in portraying the animal in a way that science doesn't.

The fat *T. rex* standing tall is a fantasy based on outdated science and art. Toymakers and moviemakers may not yet know better. Or even if they do know a more accurate way to show *T. rex*, they'd still rather give us our favorite dinosaur the way we're used to seeing it.

In recent years a couple of top dinosaur artists have tried to remake *T. rex* according to how scientists now

ABOVE: *T. REX* SCENE IS JUST ONE DETAIL FROM RUDOLPH F. ZALLINGER'S EPIC *THE AGE OF REPTILES* MURAL FOR THE PEABODY MUSEUM OF NATURAL HISTORY AT YALE UNIVERSITY.

ABOVE RIGHT: *T. REX* WAS STILL LOOKING LIKE A TAIL-DRAGGING, ERECT PORKER WHEN THE LOUIS PAUL JONAS STUDIOS BUILT THIS MODEL FOR THE 1960 NEW YORK WORLD'S FAIR. THIS CAST IS ON DISPLAY AT THE BOSTON MUSEUM OF SCIENCE.

RIGHT: *T. REX* HAD THREE FINGERS AND SOME TOUGH LUCK IN *VALLEY OF GWANGI*, A HORROR MOVIE IN MORE WAYS THAN ONE. *T. REX* WAS STILL SEEN AS A STAND-UP BEHEMOTH IN THE 1950s.

see it. A pioneer in the overhaul of *T. rex*'s image was sculptor and exhibit builder Richard Rush of Chicago. With Princeton University paleontologist Don Baird as scientific advisor, Rush determined that "the typical pose of *T. rex* rearing up on its hind feet probably never took place." Instead, Rush posed *T. rex* "head down, tail sticking out for balance." Rush's dynamic thirty-four-foot-long *T. rex*es stand in museums in Indianapolis, Greensboro, and Milwaukee.

These days the big *T. rex*es most folks see are robots. The scaled-down robot *T. rex*es built by Dinamation and Kokoro (a Japanese company I'm an advisor to) and rented out to museums are built along the lines of how scientists now see these animals—tail up and head down. They move around a little, growl, and do a great job of scaring little kids.

But the best *T. rex* since the real thing is the life-size robotic one that animator Phil Tippett designed and special effects genius Stan Winston built for *Jurassic Park*, the Steven Spielberg movie of Michael Crichton's book. The book is about dinosaurs come back to life, and this *T. rex* comes darn close. The producers asked me out to the set a few times as a scientific advisor to the film. The *T. rex* is sleek and really scary. Not only is it a great-looking *T. rex*, but they've figured out ways to make it move that don't look clunky at all. If you somehow miss the movie, you'll probably be able to see this *T. rex* at a Universal Studios theme park.

Pose the *Jurassic Park T. rex*, or even one of Rush's *T. rex*es next to one of the museum paintings of *T. rex* from fifty years ago and I think you'd have a neat display of our changing view of *T. rex*. Of course, no two scientists or artists imagine *T. rex* quite the same way, so the best modern *T. rex*es don't look alike. They all start from the same skeletal parts. But they don't end up the same. And they usually don't end up as skeletons at all.

I'm not partial to museum skeletons myself. It's not that I don't like skeletons; it's mounted ones I don't like. Often, mounts are discovered to be wrong not long after they're made. As soon as we take a skeleton out of the

LEFT: A FULL-SIZED AND AMAZINGLY MOBILE *T. REX* WAS FASHIONED AT STAN WINSTON STUDIOS FOR THE MOVIE *JURASSIC PARK*.

RIGHT: I'M AN ADVISOR TO KOKORO, THE JAPANESE ROBOTICS MANUFACTURER THAT CRAFTED THIS *T. REX*.

BELOW: *T. REX* IS MORE ACTIVE AND HORIZONTAL IN THIS 1976 MODEL BUILT BY THE RICHARD RUSH STUDIOS UNDER THE SUPERVISION OF MY BOSS AT THE TIME, DR. DONALD BAIRD OF PRINCETON UNIVERSITY.

ground and reposition it, we're on the trail of fantasy.

And I think skeletons on display make museums into halls of death. Our Museum of the Rockies dinosaur hall was finished a few years ago, and it doesn't have a single dinosaur skeleton. You go to one of those big eastern museums and you're looking at just a bunch of skeletons that have been stood up—dead animals in a standing position. I think that's a peculiar way to look at dinosaurs. And I think it's part of the reason most people think of dinosaur paleontologists as guys who just dig up bones and stick them together.

I don't know what we should do to remake our image as scientists. Maybe we should be called dinosaur paleobiologists instead. As for *T. rex*, scientists have begun remaking its image with the bones themselves. In the last decade a couple of museums have bought fiberglass casts of the bones from the *T. rex* mount at the American Museum in New York and have repositioned them for their own displays.

The "new" *T. rex* is a star attraction at the Philadelphia Academy of Sciences' excellent dinosaur hall. Ken Carpenter, who has studied tyrannosaurs and worked for the Museum of the Rockies and is now working as a preparator at the Denver Museum of Natural History, did this *T. rex*. It's a great mount, very believable. *T. rex* is nearly horizontal, striding with its tail up (minus several of the American Museum's extra vertebrae) and mouth open. The folks at the Royal Tyrrell Museum of Palaeontology in Canada used a similar pose when they positioned the cast bones from the *T. rex* they excavated at Huxley (with a cast of the American Museum skull).

Colorado dinosaur paleontologist Bob Bakker reconstructed *T. rex* differently at the Denver Museum of Natural History. With the lightweight hollow fiberglass casts of the American Museum *T. rex* and a few iron supports, Bob was able to bring Osborn's dream of a highly dynamic *T. rex* mount to life. In Denver, *T. rex* is kicking up one leg like a football punter, its head turned to the side to menace visitors as they enter its hall. It's an exciting stance. (Peter May did a similar mount of *T. rex* for the visitors' center set of *Jurassic Park*.) But looking at a kicking *T. rex*, I agreed with Peter Dodson, a University of Pennsylvania dinosaur scientist who has studied dinosaur mechanics. Peter said, "It's great art, but not great science." It's unlikely to both of us that a several-ton animal could stand on one foot and kick.

The American Museum's *T. rex* is going to be a completely different-looking animal when it's remounted for the 1994 reopening of the dinosaur halls. Mark Norell, an American Museum curator and paleontologist, told me one idea museum staff considered was to pose two tyrannosaurs, one in the original upright stance and another stalking. This idea was voted down in favor of just the stalker, a *T. rex* with its head down and its back horizontal. The curators and exhibit designers also wanted the *T. rex* posed with its mouth shut, instead of with the gaping jaws you see on other *T. rex*es. They got their wish there, though it will be mounted next to a *T. rex* skull with its mouth open.

We'll be mounting our own *T. rex* display in 1993, thanks to the support of a Japanese museum. We'll have two *T. rex*es crouching over a dinosaur corpse. It will be different from any *T. rex* display before, except perhaps for one Osborn had in mind.

Reared back, leaning forward, or high kicking—

they're all *T. rex*. You can force casts into any position. And for a moment at least, *T. rex* might have been able to lean back on its tail, or raise one leg high. But if you ask me which pose is the most likely, the one we might see *T. rex* in if we ran into it on the street, I'd say Ken Carpenter's forward-leaning *T. rex* in Philadelphia.

Building skeletal replicas is an expensive business, and I'm not sure it's worth the expense. For one, I'd never use the bones themselves. They're too delicate and important for study to have them glued and drilled and put out of reach.

And I don't think people appreciate the expense involved in mounting a dinosaur skeleton. It can cost tens of thousands of dollars to buy casts of the bones from the American Museum, more if you want to prepare and cast your own museum's bones as we'd do.

Then you've got to put them all together. At the Tyrrell Museum it took a crew of five three years to prepare and mount the *T. rex*. That's several hundred thousand dollars in labor. When the American Museum finishes fixing up its *T. rex* in a more modern pose, and putting it and the museum's other vertebrate fossils in new halls, the bill will run to about $40 million! And they'll have some fleshed-out models, not just skeletons.

One great value of these huge skeletons is that when they're done right, it gives all of us, and especially the artists who depict *T. rex*, a good jumping-off point for imagining *T. rex* in the flesh. When it comes to putting flesh and scales on the bones of a *T. rex*, we can really get to speculating. No one knows what color a dinosaur was. But if we've got to guess—for museum restorations, models, and illustrations—what color do we choose? Green has been a longtime favorite since many living reptiles are green, and dinosaurs were long imagined as lumbering cold-blooded reptiles. Lately, artists who want to depict agile, hot-blooded, birdlike dinosaurs have gone to the opposite extreme.

Now dinosaurs are as bright as overgrown parrots in a lot of illustrations. If you look at living animals today, most blend into their environment. That's important for predator and prey alike. The biggest animals, however, come in solid and drab colors, from whales to elephants. Of course, these animals are mammals, and most mammals are color-blind. Dinosaurs' nearest relatives, birds, can be colorful, though ground-dwelling birds are less bright than canopy fliers. If *T. rex* blended into the forests and well-vegetated river deltas of its time, it might have been a mottled green and brown. Its close cousin *Tarbosaurus* lived in drier territory, in what is now the Gobi Desert of Central Asia. It might have been decorated in dustier, paler tones.

As far as we know, any coloration was possible for *T. rex*. I just happen to prefer mine in conservative colors.

As for what the skin looked like, we have a better idea. *T. rex* is customarily given a scaly hide made up of many small bumps. That's a reasonable assumption because

we have skin impressions from several dinosaurs. The nearest predator in size and shape to *T. rex* for which we have skin is *Albertosaurus*, and its skin was packed with small bumps. It is possible these raised areas were display features. That means they might have been colored all the time, or sometimes flushed with color to attract a mate or to show dominance to a rival.

Stephen Czerkas, a leading sculptor of dinosaurs for museum exhibits, takes his own dinosaur studies very seriously. When he built his first full-sized dinosaur in his garage, a twenty-five-foot *Allosaurus*, he copied a skin pattern of keel-shaped rosettes from the mummy of a duckbilled dinosaur because no skin was known from any meat-eating dinosaurs (we've now got some from an *Albertosaurus* at our museum). Steve hand-stamped that pattern on his clay sculpture scale by scale. When he heard of a *Carnotaurus* skin discovery by Argentine scientist José Bonaparte, Czerkas erased his entire pattern. He not only redid the skin of his allosaur, he went to Argentina and helped discover more *Carnotaurus* skin impressions with Bonaparte.

Czerkas adapted those *Carnotaurus* skin patterns when he designed a full-scale fiberglass *T. rex* replica that is now displayed in an entire dinosaur hall of Czerkas creations at Taipei's Natural History and Dinosaur Museum in Taiwan. Czerkas's Taipei *T. rex* is a muscular animal. Its head is turned, and its body is leaning forward, but its feet are planted squarely on the ground. Czerkas has colored it gray-green with brownish banding. That's colorful by Czerkas's standards.

The best modern dinosaur painters have gone further in coloring *T. rex*. John Gurche, Doug Henderson, Brian Franczak, and Greg Paul are among the top dinosaur artists who've painted *T. rex*. All of them have made *T. rex* into a far slimmer, more graceful-looking animal than Knight and Zallinger did. Of today's top dinosaur painters, Mark Hallett has been the boldest in coloring *T. rex*. Before painting *T. rex*, Mark consulted with experts—Ralph Molnar on *T. rex* anatomy, Steve Czerkas about dinosaur skin patterns. Then he painted *T. rex* in green with magenta and brown stripes, a giant camouflage suit. He says, "They probably needed a pattern to break up their huge body size, like a tank on maneuvers, to hide from their prey." While that's not how I imagine *T. rex*, it is certainly possible.

All of these artists are very diligent about doing research on their subjects. They consult closely with paleontologists. But their conclusions and their images of *T. rex* vary, in more ways than their colors. Czerkas may be conservative about coloring dinosaurs, but he is more daring in depicting *T. rex*'s skull anatomy. He gives it bulbous pointed horns around the eyes and nose. There are few indications for all these features on the skull of a *T. rex*.

But I can see adding small hornlike structures around the eyes of *T. rex*, and my former boss at Princeton, paleontologist Don Baird, agrees. When he looked at a top view of a cast of the American Museum *T. rex* skull, he noticed a rugosity, or bumpy patch, around the eyes. Don says, "It's clear to me *T. rex* had horns over its eyes." (Don made his conclusion after he'd advised Richard Rush on Rush's first *T. rex* sculpture, so Rush retrofit his dinosaur with a bump.)

Bony protrusions are found on earlier dinosaur predators like *Carnotaurus* and *Allosaurus*. However, sometimes animals have horns for which we find no evidence on their skeletons. Lots of horns can be seen on the head of Steve Czerkas's pet rhinoceros iguana (named Don, as in *Iguanodon*), which he uses as a model for his sculptures. "It has several horns, but indications of only one on its skull," says Czerkas. Paleontologist and artist Bob Bakker thinks *T. rex* (and the pygmy tyrannosaur he, Phil Currie, and Mike Williams of the Cleveland Museum of Natural History identified) had bumps over its eyes, not horns. And those bumps were for head-butting (see Chapter 6). As with many behaviors imagined for *T. rex*, calling it a head-butter seems like too much guesswork to me.

Paintings often show *T. rex* doing things I don't think it ever did. Mark Hallett has painted two *T. rex*es

RIGHT: SPECIAL-EFFECTS CREATOR JIM DANFORTH MADE THIS HUNGRY *T. REX* FOR THE MOVIE *CAVEMAN*.

BELOW: 1930 WORLD'S FAIR SINCLAIR EXHIBIT DREW MILLIONS TO SEE *T. REX* AND FRIENDS.

wheeling menacingly in the dust around a herd of *Triceratops*es. The horned dinosaurs are lined up, horns outward, in a defensive ring, sheltering their young in the center. To me, it looks like pioneers putting the wagons in a circle when Indians attacked. I think it's unlikely it could have happened that way. *T. rex* wouldn't waste its energy attacking a herd, and I don't think a bunch of pea-brained *Triceratops*es would get together so neatly. There's absolutely no evidence for cooperative hunting in *T. rex* or cooperative defense among horned dinosaurs. Mark says, "Paleontologists, Bakker among them, have suggested they might do this. Musk oxen do, and dinosaurs might have done many things like modern mammals."

HERE'S THE WAY I THINK *T. REX* REALLY LOOKED. SCULPTOR MATT SMITH MADE THIS MODEL BY STUDYING KATHY WANKEL'S *T. REX* SKELETON, THEN CAREFULLY SCULPTING MUSCLE AND SKIN. THIS *T. REX* LOOKS A LOT LEANER, IF NOT MEANER, THAN IT'S EVER BEEN BEFORE.

Maybe. But it seems to me that we've bounced from one extreme to the other in how we depict *T. rex*. If *T. rex* isn't the drab, lumbering lummox I grew up with anymore, it's now a giant killer ballerina in a tutu, the flashy animal Osborn first wanted it to be.

I'm not interested in dancing dinosaurs. I'd like to see a *T. rex* sculpted or painted accurately according to what we now know of how its bones looked and usually moved.

Sculptor Matt Smith has done just that. Matt made his *T. rex* by looking at the bones of an animal he'd helped dig out of the ground and clean in our lab. Matt's dug and researched our *T. rex*, and prepared other fossils, and made many replicas for the Museum of the Rockies and other museums. (He and Ken Carpenter did an eye-opening study of the arm motion and power of *T. rex*. See Chapter 5.) Matt's made a *T. rex* model based as closely as anything yet done on the most current fossil evidence. The coloring is Matt's preference, but the sculpture is modeled on what the bones and the insertion marks of the muscles suggest this animal could do. His *T. rex* doesn't leap around like the Karate Kid, but it is agile and hostile, and real enough to scare the hell out of me. And most importantly for all of us, it is real, because Matt's *T. rex* is based on fossil evidence, all of Matt's art and our science on *T. rex* are based on what we do with the bones we dig out of the ground.

CHAPTER 5

THE BARE BONES

NOW THAT WE'VE GOT all the bones back in the lab, what do we do with them? Usually we put them in storage for a few years. Unlike wine, they don't get any better in the basement. But, like other paleontology museums, the Museum of the Rockies has a huge backlog of fossils waiting to be cleaned. And just because *T. rex* has the biggest fan club doesn't mean we can shove aside the preparation of other valuable study specimens when a *T. rex* comes in the door.

We've got a waiting list for fossil cleanup partly because we're good at finding dinosaurs and other fossil animals. We don't prepare just any old bones, only those that are likely to offer new information about the animal, or are part of an unusually well preserved or complete skeleton. We've recently found a lot of those, especially dinosaurs.

Also, like most paleontology museums, we can't afford the trained preparators needed to clean our fossils. I already mentioned the expense of preparing dinosaurs and the time needed—thirty thousand hours, or five people working for three years, for the Royal Tyrrell crew to clean a single *T. rex*. The Tyrrell used to have twenty preparators, but with economic slow-downs, they're down to a handful. We can afford only a couple of paid workers on the money we raise. Fortunately, we also have some top-notch volunteers. Other museums are worse off. Brigham Young University has one of the world's largest collections of dinosaurs. Most of it has been in plaster jackets for two decades, and the huge bundles fill the basement of the bleachers of their giant football stadium. "Dinosaur" Jim Jensen collected some of the world's biggest dino-

saurs for BYU, and nobody's quite sure if there isn't a
bigger one still in those jackets. But who knows when
we'll find out, since BYU had to let its last paid preparator
go this year.

Kathy Wankel's *T. rex* was bound to get worked on
pretty fast, though, at our museum. For one reason,
unless we worked on it, we couldn't even get it inside
the door.

When Bill McKamey finished driving Kathy's *T. rex*
across Montana to Bozeman on July 3, 1990, he pulled
up to the back of the museum, where the staff was ready
with a forklift to haul the bundles on pallets into the
loading dock. We knew *T. rex* wasn't going to make it
into the lab the way it looked. The big bundles were so
heavy they exceeded the architect's estimates of the
carrying capacity of the museum's floors. Moved just as
they were, the bundles would have broken through the
floor of the museum.

We opened up the plaster jackets and removed
sediment from the hip and vertebrae bundles on the

THE WANKEL *T. REX* GETS
DROPPED OFF AT THE
MUSEUM OF THE ROCKIES.
SOME OF THE BONE BUNDLES
WERE SO HEAVY THAT THEY
WOULD HAVE FALLEN
THROUGH THE FLOOR IF WE
DIDN'T TAKE THEM APART
FIRST ON THE LOADING DOCK.

THE PREPARATION OF THE
WANKEL *T. REX* GOES ON IN
FRONT OF THE PUBLIC AT THE
MUSEUM OF THE ROCKIES. WE
EXPECT TO HAVE THE
SKELETON CLEANED AND CAST
BY JUNE OF 1993. THAT'S
THREE YEARS AFTER WE DUG
IT UP, NOT LONG BY
PALEONTOLOGY STANDARDS.

loading dock in a matter of hours, making them into smaller units. Instead of hauling the jackets into the storage area or the preparation laboratory next to my office in the basement, we brought them all upstairs to the museum display area, where we set up a special room to house the *T. rex* while three of our volunteer preparators (and sometimes Pat Leiggi) are working on it. The jackets are on tables behind big glass windows, and on the far wall of the big room is a mural of a *T. rex* skeleton. As visitors walk by, they can see the preparators working on the bones.

Fossil preparation in the lab is a difficult art. You've got to know anatomy and how to handle small tools, more delicate ones than you use in the field. You need a soft touch. And above all you need unbelievable patience. I prepared fossils for years at Princeton before I became a paleontologist, and as long as someone's paying me to look for and study dinosaurs, I don't want to be a preparator (though I still prepare really delicate and important fossils in my work, such as the embryos I've found in dinosaur eggs).

The equipment isn't anything too complicated. The fanciest tool we use is an air scribe. It's like a miniature jackhammer with a point as small as a pen. The tip vibrates in and out, chipping away at the rock surrounding a fossil. We couldn't use air scribes on *T. rex* since the sandstone around the bones was too soft. Instead we used dental tools and paintbrushes and sometimes a toothbrush to clean the bones.

Each *T. rex* bone is prepared separately. Since all of us wanted to see the skull most of all, we worked on those bones first, except for the arm that preparators Ken Carpenter and Matt Smith had begun work on even before we excavated the rest of *T. rex*. We left the huge hip jacket for last because it needed so much work.

The bones were cleaned individually. If we found a crack in the bone, we removed the rock that had accumulated in the crack. If the remaining pieces of bone fit together, we glued the fragments. If there wasn't a clean fit, we didn't try to force things. Instead, we kept the parts, always logging in a collection number and a description on a tag tied to each bone.

As I'm writing, it's been more than two years since we excavated our *T. rex*, and we've still got at least a year to go before we're done cleaning it. The skull bones have all been prepared, and they're laid out—brown, shiny, and smooth—along with some of the arm, leg, back, and tailbones on long open-sided metal cabinets in our basement. That way I, or a visiting scientist, can just walk over to them, pick them up, measure them, and study them anytime we want. They're so much fun, I sometimes go into the collections just to look at them.

And we still don't know exactly how much of *T. rex* we've got. Until we've cleaned up all the fossil bundles, we won't know if we've got, say, the bones of the other arm of *T. rex*. We're pretty sure we're missing the back end of the tail, some jawbones, one arm, and half of the rib cage. But we've still got a lot of *T. rex* parts to look at.

Aside from pleasure, what do we get from these bones? A lot. Individually, and together, they can tell us about *T. rex*'s evolution, its movements, its five senses, its behavior. By looking at specific bones and groups of bones, we're learning a lot more about how *T. rex* might

PREPARATOR AND ARTIST KIT MATHER IS AT WORK HERE ON *T. REX*'S SKULL.

have looked and functioned. I'll describe some implications these findings may have for *T. rex*'s behavior in Chapter 8. For now, here's some of what we're finding as we're cleaning specific areas of *T. rex*'s skeleton.

THE SKULL

The skull is the most complicated piece of any animal's anatomy, and the one that undergoes the most evolutionary change. It has many openings (fenestrae) and tubes for air passages and nerve canals. *T. rex* had an enormous skull, nearly as long as I am, and I'm six feet tall.

Where it is solid, the roof of *T. rex*'s skull is three or four inches thick—"heavy duty," in the words of paleontologist Phil Currie. On some dinosaurs, skulls were light, small, and delicate and so very hard to come by as fossils. The biggest dinosaurs, such as the four-legged sauropods *Diplodocus* and *Brachiosaurus*, had tiny skulls with brains no bigger than those of house cats. We know hundreds of sauropod skeletons, but in two centuries, paleontologists have found fewer than two dozen sauropod skulls.

For horned dinosaurs, such as *Triceratops*, the preservation story is the opposite—we've got lots of skulls but relatively little of the rest of the body. But you can get lucky. Out in the Hell Creek badlands in the summer of 1991, Diane Gabriel, a graduate student at our museum, found the skull and most of the skeleton of a *Triceratops*.

Our *T. rex*'s skull is big and sturdy, and though full of holes, it seems to have held up pretty well. Phil Currie and I will look closely at the skull and at the brain cavity in particular. And thanks to the folks at General Electric in Cincinnati, we're now able to look at *T. rex*'s skull inside and out in thousands of ways.

So much of the interesting anatomy of a skull is inside it—in the braincase, air passages, and nerve pathways. It's hard to get a good look at these features without

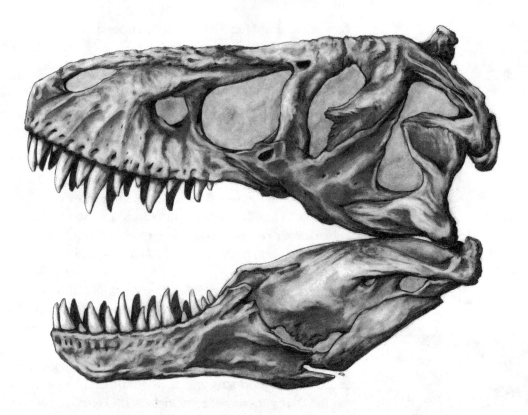

breaking apart the individual skull bones. Then it's difficult to get them back together. Breaking them apart is not a risk you'd want to take with the few *T. rex* skulls in existence. But with a CAT scan you can photograph thousands of microscopically thin sections of bone and then piece the images back together in a three-dimensional view from any direction. And since it's all done by X rays, you never damage the skull.

Our *T. rex*'s skull is too big to fit in hospital CAT scanners, but at the GE labs in Cincinnati there's a CAT scanner big enough to handle a five-foot-long skull. We trucked *T. rex* out there in a U-Haul, along with every other good skull we had in our collections—two and a half days of nonstop driving. (As the boss, I got to take the plane.)

We took thousands of pictures of *T. rex*. It took us seven hours just to photograph the *T. rex* jaw, since we made an image every fiftieth of an inch along the skull. In the end we had a gigabyte's worth of image informa-

THE SKULL OF *T. REX* IS A GREAT MAZE OF BONE AND TUNNELS.

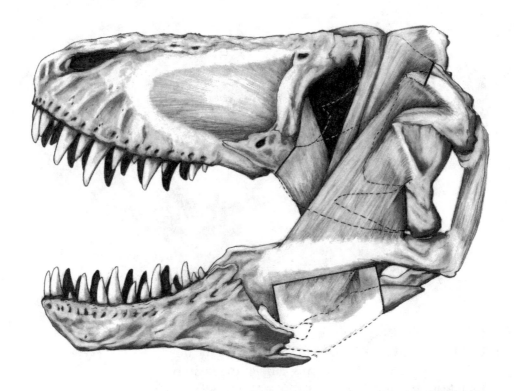

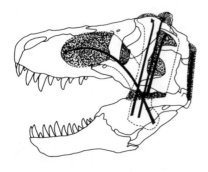

STUDY THE ORIGIN AND INSERTION POINTS ON THE SKULL AND YOU CAN MAKE A REASONABLE GUESS AT WHAT *T. REX*'S SKULL MUSCLES MIGHT HAVE LOOKED LIKE.

tion—that's a thousand megabytes. GE says they'll let us use their Cray computer, one of those super number-crunchers, to help generate and analyze our images. Sue, the Black Hills Institute *T. rex*, was scheduled for a similar CAT scan treatment in spring of 1992 at the Marshall Space Flight Center in Alabama, just days before the skeleton was impounded. It was never scanned.

It's hard to know what we'll find, even what questions to ask, until we get a look inside the skull with CAT scans. We need to see how thin and how porous the bones are. We'll be looking at tooth growth, upper and lower jaw mechanics, the sinuses, and the size and shape of the braincase. CAT scans of *T. rex*es of different ages would be a great source of information. It would be interesting to look at changes in the brain, shown in the braincase, as the animal grows up. We can't yet do that with *T. rex*, since we have only adult animals.

Paleontologist Ralph Molnar has done the most de-

tailed studies of *T. rex*'s skull. In fact, they are the only scientific studies since Osborn's early in the century. Each feature of the *T. rex* skull that Molnar outlines in such detail helps determine which aspects of *T. rex* distinguish it from other dinosaurs. For instance, Ralph notes fifteen skull features that distinguish *T. rex* from its closest relatives, the other tyrannosaurids. One is beady eyes—*T. rex* had a narrow eye socket with what San Francisco paleontologist Jacques Gauthier calls a narrow "keyhole" shape. The eye sockets of *Alberto-saurus* and other close relations of *T. rex* were oval or elliptical.

A key similarity between the skulls of birds and the late dinosaur carnivores, big and small, is the many holes in their skulls. These holes could have served any number of purposes, and in *T. rex* the most logical is the simple one of reducing the weight of the huge skull. Among the many holes in *T. rex*'s head were lots of air spaces to help keep the skull light and make room for an elaborate network of muscles, blood vessels, and nerves.

There were other big holes in the *T. rex* skull that may

PUT FLESH OVER THE MUSCLE AND YOU HAVE A PRETTY LIFELIKE *T. REX* HEAD.

T. REX COULD SPREAD ITS JAW WIDE, THOUGH IT COULDN'T EXPAND ITS SKULL LIKE A SNAKE'S.

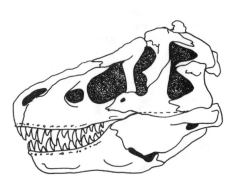

T. REX COULD CLOSE ITS MOUTH WITH GREAT FORCE AND EFFECT, THANKS TO POWERFUL JAWS AND HUGE, SHARP TEETH.

THE NASAL BONE OF T. REX IS A PLATE STUDDED WITH PECULIAR HOLES.

be sinuses like ours. One system of *T. rex* skull holes connected to the nose, another to the middle ear, and another led from the back of the skull to air sacs in the lungs. At least the tubes leading to the lungs would have been filled with air. There were also many holes for nerves in the snout of *T. rex*.

The nasal bone of *T. rex* is an especially fun bone. It's a long thin strip, like the bark off one side of a log. And it's rugose—full of bumps).

Besides its many holes, *T. rex*'s braincase had room for a brain that was one of the largest in the history of life, bigger than a gorilla's or a chimpanzee's, though not as big as ours.

We can think up all kinds of talents for *T. rex* from guesswork about its skull. Maybe it could hear and see with depth perception and feel with lips. But the holes in *T. rex*'s skull don't prove very much about *T. rex*'s senses.

What we can tell with more certainty from examining many of *T. rex*'s skull features is how it ate. *T. rex*'s wide cheeks and deep jaws supported bigger, more powerful jaw muscles than those of any other dinosaur we know. Atop its head, between the temples, was a crest of bone. Muscles along that ridge, and extra long muscles around the lower jaw, could give *T. rex* a powerful bite.

T. rex may have had an overbite, as skull expert Ralph Molnar contends. I don't know of any tyrannosaur jaws that shut all the way. So is the overbite we see natural, or an after-death change in the jaw alignment? According to Ralph, it's natural. With its powerful cheek muscles and overhanging top teeth, *T. rex* could have

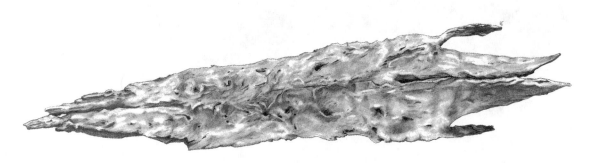

combined two different biting techniques: scissors and the nutcracker bites. Scissoring could shred meat, nutcracking could chomp into bone. But at least one study indicates *T. rex*'s bite wasn't so efficient for clean bites and instant kills.

Reservations about *T. rex*'s chewing style shouldn't take away too much from *T. rex*'s impressive skull power. *T. rex* had neck muscles a football lineman would envy. Its skull bones had thick walls and were joined more tightly than in other hunters. Bars across the eye and cheek sockets added more stability. Bigger teeth and fewer of them, relative to other tyrannosaurs, also contributed to *T. rex*'s biting force.

T. rex's palate was light, supported by thin struts that would have allowed the skull to flatten out sideways with the force of a bite, and stretch a bit to help hundreds of pounds of dinosaur tissue slide down its gullet.

That doesn't mean *T. rex*'s jaws could spread open like a snake's. The right

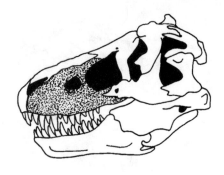

MAXILLA: *T. REX* HAD HUGE MAXILLARY BONES, AND IT NEEDED THEM TO ACCOMODATE ITS ENORMOUS TEETH. THE MAXILLA IS THE MOST STURDY BONE IN THE HEAD. THE HUGE MUSCLES WHICH ATTACHED TO IT PROVIDED THE POWER TO CRUSH AND CHEW *T. REX*'S FOOD. THE SIZE OF THE TOOTH SOCKETS IS ESPECIALLY IMPRESSIVE.

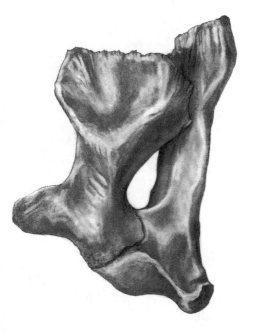

and left jawbones were joined at the chin like ours, but not fused together. Maybe they slid back and forth against each other slightly, but we don't know that. The back of the lower jaw was firmly joined to the skull by a bone called the quadrate. Jaw bones by the bottom of it pivoted on a hinge, allowing *T. rex* to spread its jaws and open its maw wide. How wide we don't know. A third joint allowed flexing between the toothy front and the muscular back of each jawbone. Put it all together and you have a jaw anatomy with aspects of the strength of a crocodile and the lightness of a bird.

With this equipment, it's not likely *T. rex* was a finicky eater. But it could have chewed well before swallowing. Sure, it could have eaten Arnold Schwarzenegger in one bite or turned him into two hors d'oeuvres with a single snap. But there's no reason to swallow a lump that big when you could take it apart first with a spectacular set of teeth.

TEETH

T. rex had bigger choppers, more varied in size and shape, than those of other tyrannosaurids. There might have been fifty teeth in its mouth at any one time. Up front, it had incisor-like teeth, as all tyrannosaurids do, four on each premaxilla, the bone in the front of the upper jaw. Among meat-eating dinosaurs, only the little *Troodon*s and tyrannosaurids had these. The shape of these teeth was fairly wide and flat, designed for nipping, according to tooth expert Phil Currie. They would have been more precise in raking meat from bone, Phil says, without biting right through the bone as *T. rex*'s bigger teeth could.

T. rex had twelve or thirteen additional teeth on each side of the upper jaw. These cheek teeth were enormous, a foot long including the root, and as big around as a child's fist. They were some of the biggest teeth of any dinosaur (*Spinosaurus*, a North African dinosaur known only from bits and pieces, had bigger teeth and may have been a bigger carnivore). You'd expect *T. rex*

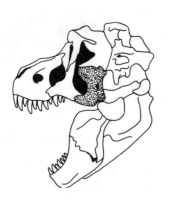

ABOVE: THE QUADRATE.

BELOW: THE QUADRATE IS PART OF THE HINGE THAT WORKS *T. REX*'S JAW.

to have big teeth, since it was one of the biggest carnivores ever. But as Jim Farlow, a dinosaur paleontologist at the University of Indiana, Purdue, found, even in proportion to *T. rex*'s size advantage over other predators, its teeth were oversized.

With thick, strong teeth the size of bananas, *T. rex* could well have penetrated the hide (except when armored), the flesh, and the bones of another dinosaur with a single bite. We know *T. rex* broke off some of those teeth on or in the bones of other dinosaurs.

Fortunately for *T. rex*, like all dinosaurs, it was always making new teeth. Whether or not they broke off in feeding, old teeth were constantly being replaced by new teeth that pushed their way out of every socket. Within the jaw, two or three teeth for each tooth position were always developing, getting ready to emerge.

Greg Erickson, a graduate student of mine, has made a special study of our *T. rex*'s teeth. Greg looked at what are called "von Ebner lines," incremental markers of growth in the dentin at the edge of pulp cavity of the tooth. Greg compares these lines to "reverse tree rings. They grow from outside to inside." You can see these thin lines under a microscope if you cut a thin section of a tooth and look at it at 100× magnification. Greg figures that an adult *T. rex* would have laid down one brown and one black line each day.

We can't know for sure how fast *T. rex* made those tooth rings. But Greg found a clever way to test for an answer. He went to a crocodile farm in Louisiana and fed tetracycline to the crocodiles. The antibiotic has the neat effect of staining their day's tooth growth chartreuse green. Greg also injected Puff and Bubba, two small crocodiles we keep at the museum. None of his subjects was happy about a needle in the belly or the base of the tail, and Greg wasn't too eager to give it to them, but he had help from some experienced crocodile-handlers.

Days later, after the farm crocodiles were on their way to becoming shoes and handbags, Greg got their three-

THIS CLAWLIKE FOSSIL IS A
TOOTH FROM "STAN"
THE *T. REX*.

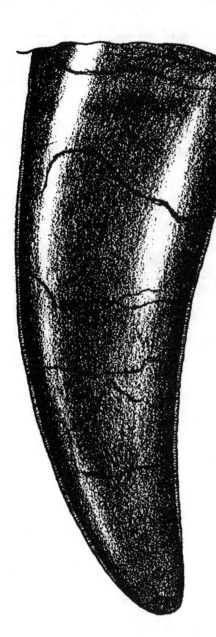

A *T. REX* TOOTH,
ACTUAL SIZE.

inch-long teeth. He sectioned them and found that the dye was confined to a single set of growth rings. It's a reasonable guess that dinosaur teeth grew new rings daily as well. My fellow vertebrate paleontologists in North America thought well enough of Greg's study that they made him a co-winner of the 1991 student prize for the best presentation at our annual meeting.

Counting up the number of lines on *T. rex* teeth, Greg worked out that *T. rex* would have shed its teeth every two or three years. Crocodile teeth don't last much more than a year. And Greg figures a duckbill tooth would grow in within half a year.

T. rex had such huge teeth that you'd figure it would need a while to grow them. Most dinosaurs were always producing backup teeth. But *T. rex*'s replacement teeth didn't start growing until its working teeth were 70 percent grown. Crocodiles start building replacements when their working teeth are less than half-grown.

T. rex had a never-ending supply of teeth. But if its teeth broke off often, *T. rex* would have had an uneven bite. Then it may have been hard for it to make a smooth, slicing cut. For a replacement tooth to fully emerge might take eight months to more than a year. That's how it is with crocodiles, which also replace their teeth continually throughout life.

T. rex's teeth were serrated, so they could saw as well as puncture and rip meat. The serrations were extremely wide, with razor-sharp edges, "beautifully adapted to saw bone and meat," according to Phil.

The back of the *T. rex* tooth had the biggest serrations, so perhaps *T. rex* bit down hard to break skin or bone, then cut backward through its prey. Serrations might have been a way to get the cutting action of a thin blade without sacrificing the strength of a stout tooth. Jim Farlow, an expert on dinosaur teeth (and footprints), thinks that's true. But Jim thinks the small serrations might have helped to cut by binding tissue more than by slicing it. Coupled with *T. rex*'s enormous biting strength, the binding would have resulted in more tearing and a better grip on the victim.

Jim is impressed that *T. rex*'s upper teeth combined features that made them useful for cutting into and grabbing onto prey with other features that would make them withstand the stress. Each tooth had a good point. It wasn't blunt at the tip like the teeth of animals that crunch shells or grind vegetation. It was knife tipped to pierce flesh, maybe even break through bone. Overall, the tooth was shaped somewhat like a blade, though not so flat. And the base was fairly broad, providing resistance to lateral forces that might have made a slimmer tooth break off.

Jim has studied more than five hundred meat-eating-dinosaur teeth and calculated their bending strength. He finds tyrannosaur teeth superior in strength to those of any other carnivorous dinosaur. Maybe they were thick just because of their size, but that thickness would have helped them bite into, even break through, solid bone. It's hard to say exactly what *T. rex* teeth could do, as no modern carnivores I know of, except sharks and Komodo dragons, have serrated teeth. Modern mammalian hunters use their big canines to hold prey rather than to cut meat off their victims.

Jim and others have found that the teeth of large meat-

ABOVE AND RIGHT: *T. REX*'S TEETH WERE EVENLY NOTCHED WITH GOOD-SIZED SERRATIONS.

GREG ERICKSON FOUND THESE LINES OF GROWTH, PROBABLY LAID DOWN DAILY, IN *T. REX*'S TEETH.

eating dinosaurs were far less specialized than those of mammals—no molars or canines as you and I have. But not all *T. rex* teeth were as big and thick as bananas. Those in the back of the jaw were thinner and shorter, perhaps for fine slicing-and-dicing of bones and flesh, as paleontologist Bob Bakker suggests.

ARMS

One of the greatest features of Kathy Wankel's *T. rex* is that it has all the arm bones, shoulder to fingers. Only the claws are missing. That's the first time we've ever had all those parts from a *T. rex*. And those bones allowed my co-workers to draw some conclusions about how those arms worked, conclusions that surprised me.

What didn't surprise me was how small *T. rex*'s arms were—no longer than mine. In evolution it's an amazing reduction over earlier tyrannosaurs.

Once you look at the actual bones, it's obvious that *T. rex*'s arms were not just dangling uselessly at its side as it walked along. The arm bones have very distinct muscle scars—sometimes smooth surfaces, other times indentations in the bone made by the attachment of large muscles, or projections to which tendons attached.

T. rex's arm bones are only three feet long on an animal forty feet long. But *T. rex*'s arm bones are about three times as thick as mine.

Matt Smith, a sculptor and former fossil worker at the Museum of the Rockies, and Ken Carpenter, a researcher and fossil mounter who is now a preparator at the Denver Museum of Natural History, made a special study of our *T. Rex*'s arm bones. Ken detailed the shape of the arm bones. Matt looked at the muscle scars in the forearm of *T. rex*, the pits and grooves in the bone where muscle had attached. By measuring these scars they could estimate the size of the muscles and tendons. For instance, the bump on the humerus where *T. rex*'s biceps attached is about the size of a nickel. It doesn't sound big, but that's a lot larger attachment point than you'd see on a weightlifter's upper arm bone. The biceps didn't attach to the same point on *T. rex*'s arm as

MATT SMITH RECREATED *T. REX*'S ARM MUSCULATURE BY MEASURING THE ORIGIN AND INSERTION POINTS FOR ARM AND SHOULDER MUSCLES. EXCEPT FOR THE BICEPS WHICH CORRESPONDS IN SIZE TO THE RESULTS OF A BIOMECHANICAL STUDY MATT AND KEN CARPENTER DID, THE REST IS EDUCATED GUESSWORK. SO MATT KEPT THE MUSCLES AT THE CONSERVATIVELY SMALL END OF THE RANGE OF POSSIBILITIES.

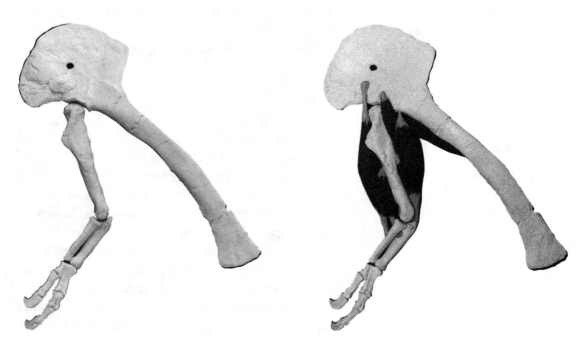

our muscles do on our arms. Our biceps attaches on our forearm, inside our elbow and along our upper arm. *T. rex*'s biceps attached farther down its forearm. That gave *T. rex* more leverage. Matt compares it to a drawbridge. If you try to lift a drawbridge with a rope, you want the rope to go out as close to the end of the bridge as possible. That way you can lift more of the bridge's weight with a given amount of force.

The muscle scars for the shoulder and upper back muscles, which on us are called the teres and the latissimus dorsi, are also massive on our *T. rex*. Emerging from a huge shoulder pad of muscle, *T. rex*'s arm would have looked even stubbier than it was.

From this kind of data, Matt and Ken worked out the range of motion that arm could move through. They also reconstructed the forelimb musculature of *T. rex* with comparative anatomy to figure out *T. rex*'s range of motion. But the comparisons weren't easy to find. No birds have forelimbs like those of *T. rex*. Birds' limbs allow for a lot of rotary movement that helps them fly.

The closest arms to *T. rex*'s among living animals belong to a crocodile. But crocodiles walk on all fours,

LEFT TO RIGHT: THE BONES OF THE ARM AND SHOULDER. THE HUMERUS IS AS MASSIVE AS THE UPPER LEG BONE OF A RACEHORSE.

THE BICEPS AND BRACHIALIS FLEX THE FOREARM, BRINGING IT UP AND IN. THE TRICEPS EXTENDS THE FOREARM.

THE SUPRASCAPULARIS AND DELTOID MOVE THE ARM OUT AND BACK WHILE THE PECTORALIS PROVIDES THE OPPOSITE MOVEMENT, ROTATING THE ARM FORWARD AND INWARD.

THE POPEYE-LIKE FOREARM EXTENSORS WORK THE HAND AND WRIST.

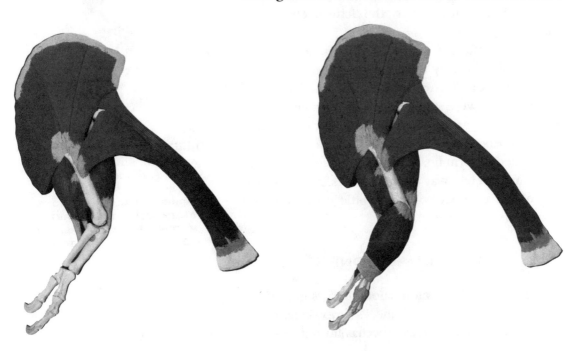

unlike birds or *T. rex*. *T. rex* arms didn't move quite like anything alive today. When Ken and Matt put muscles on *T. rex*'s forelimbs, the arms were capable of manipulating in front and back movement, and up and down motion, as well as lateral motion. But that motion was limited. The elbow end of *T. rex*'s upper arm was flat, not rounded like ours, so *T. rex*'s arm wasn't very flexible. *T. rex* couldn't extend its arm much past a ninety-degree angle between the forearm and the upper arm. Matt says no animal today has such a limited range of movement. And no other meat-eating dinosaurs, even *T. rex*'s closest relatives, the other tyrannosaurids, were so restricted in how they could move their arms.

T. rex didn't have much arm motion, but it did have a lot of arm strength. Its biceps muscle would have been six inches in diameter, three times as thick across as most of ours. Imagine our arm bones worked by muscles the size of our thigh muscles. Ken figured *T. rex* could hoist four hundred pounds toward its body at one time. On average, we can pull in only nineteen pounds. And Sue, Pete Larson's *T. rex*, had even sturdier arms. Small wonder Ken called *T. rex* "the Schwarzenegger of dinosaurs."

While those figures surprised a lot of us, Ken and Matt weren't surprised by their findings. They thought a bone as thick as *T. rex*'s upper arm could lift a lot of weight. They were surprised that *T. rex*'s lower arms were not nearly as muscular. *T. rex*'s strength, Matt says, would have been concentrated in its contracting muscles that pulled the arm up. It wouldn't have been able to hold much weight way out in its claws. As for what *T. rex* might have done with those little arms, Ken and I disagree (see Chapter 9).

INDIVIDUAL BONES

Here's a little bit of information about some of the individual bones of Kathy Wankel's *T. rex* skeleton that we've cleaned and that Kit Mather has illustrated. We've

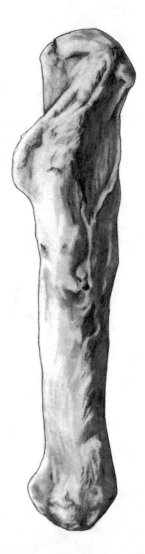

HUMERUS: THE SIZE OF THIS ARM BONE SUGGESTS IT WAS USED FOR SOMETHING, AND SO DO ITS VERY DISTINCTLY MARKED MUSCLE INSERTIONS. WE THOUGHT IT WAS PRETTY MASSIVE, BUT COMPARED TO "SUE" IT TURNS OUT TO BE RATHER LIGHT.

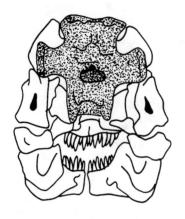

already mentioned a lot of the skull bones, so let's start at the back of the head and work our way down.

Occipital Condyle

A special swivel joint connected *T. rex*'s skull to the atlas, or first neck vertebra. This atlas vertebra fitted into a big ball on the back of the skull, the occipital condyle. On us it's just a little knob, but on *T. rex* it was the size of a small grapefruit. We have a double condyle (the ball fits into an axis vertebra on the other side), which limits the motion of our necks. *T. rex*, like birds, had only one. *T. rex*'s head was tilted forward and down when in a relaxed position. But with a single condyle, a long neck,

and powerful neck muscles, *T. rex* could swivel and look behind as well as ahead.

Neck Vertebrae and Cervical Ribs

The neck bones of *T. rex* have big prongs on them, called neural spines. These are the attachment points for huge muscles that link up on the other end to the top of *T. rex*'s head. The small size of the neck bones compared with the massive head they supported suggests that *T. rex* must have had massive neck muscles. What other benefit would those neck muscles have provided? Feel the back of your neck when you're biting down and pulling on a piece of taffy. Those muscles on a *T. rex* would have been useful for yanking at food.

Below the prong on each vertebra of *T. rex* and other dinosaurs is the rest of the neural arch. Pairs of big facets that rise vertically above the spinal cord on the front and back of each arch are called zygapophyses. They are the points at which each back vertebra is linked to the next. The space between the large zygapophyses of adjacent vertebrae on *T. rex*'s neck helped make its neck very flexible.

PAT LEIGGI TAKES A TURN AT PREPARING THE VERTEBRAE OF THE WANKEL *T. REX*.

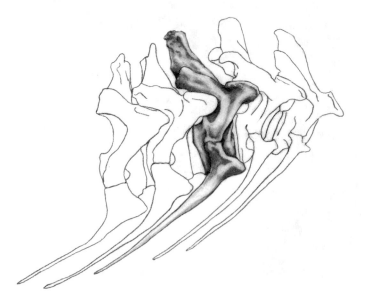

CERVICAL RIBS

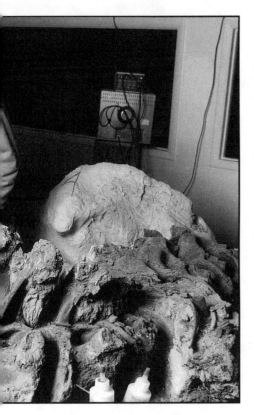

Long ribs wrap around *T. rex*'s neck. These thin bows of bone (along with the neck muscles) would have protected *T. rex*'s windpipe from attack. They were the attachment points for many muscles controlling the position and movement of *T. rex*'s neck.

Dorsal Vertebrae

The back vertebrae where the ribs attach look like coffee cans. They're huge—and they had to be to hold the big ribs coming off them and a huge stomach. Stomachs aren't preserved, but on an animal this size, a full belly could have been holding five hundred pounds of meat.

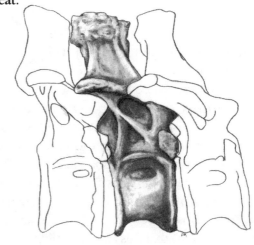

Pubis

The pubis is composed of two huge bones (left and right) that sometimes fused, with a bootlike end that hung down and forward from the pelvis. It's big enough that if *T. rex* got down on its knees, it could have rested on its pubis. I'm not saying it did sit this way, but the pubis at least makes this position possible.

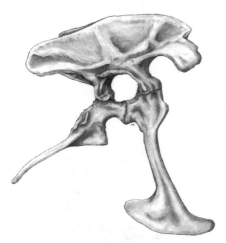

PELVIS

Ischium

A pair of these pelvc bones stick out backward. The ischia (plural) could have supported the reproductive opening, the cloaca. And they may have been a platform for attachment of big muscles of the tail.

Ilium

The third and biggest of the paired bones of the pelvis are the ilia (plural). Each ilium has a tall blade for the attachment of what might have been the biggest muscle on a *T. rex*'s body, the thigh muscle. (The femur bone fits into a big hole at the juncture of the ischium, ilium, and pelvis called the acetabulum.) This huge muscle sheet formed a triangle that attached to the femur. Similarly thick thigh muscles are one of the reasons you don't see the femur bone on a chicken—they're the thigh meat. And birds' thigh bones are short and carried horziontally. If you lived with *T. rex*, you probably couldn't have seen much of *T. rex*'s thigh bone either.

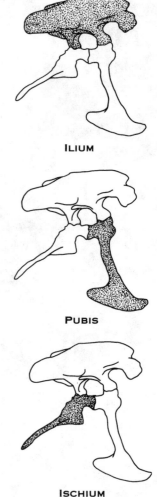

ILIUM

PUBIS

ISCHIUM

Tibia and Femur

T. rex's upper (femur) and lower (tibia) leg bones are much longer than those of far earlier meat-eaters like *Allosaurus*. *T. rex*'s svelte limbs are stubbier proportionately than those of the ostrich-mimic ornithomimid dinosaurs. Fast dinosaurs like ornithomimids had lower leg bones that were longer than those of their upper legs. That's true of other fast animals like cheetahs. But the drumstick bone, the tibia, is about the same length on *T. rex* as its thigh bone, the femur. The fact that these bones were so close in length on *T. rex* is one of the strongest indications that this animal couldn't run very fast.

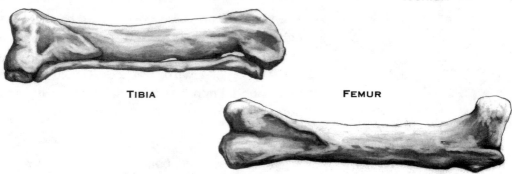

TIBIA

FEMUR

Toes

While it had only two fingers, *T. rex* had three toes (and a greatly reduced first digit)—the second, third, and fourth toes of an ancestral five-toed foot. Our five toes

and fingers are primitive characteristics in animals. When it comes to feet, dinosaurs were more advanced than we.

The middle toe is the largest and the principal weight-bearing one for *T. rex*. Like the short front teeth that it had in common with *Troodon*, the pinched third toe is a characteristic that suggests tyrannosaurids may have been descended from an ancestor of those small predators.

Tail Vertebrae

Our *T. rex* skeleton has only seventeen of what would have been at least forty-six tail vertebrae on *T. rex*. The tailbones we have are conservative in form, much like those of earlier tyrannosaurids. They're huge bones too, even bigger than those of the neck. They show projections that might have provided support to keep the tail raised horizontally, though none to limit its moving side-to-side.

Put all these bones together and you have the beginnings of an understanding of how *T. rex* looked and moved. And by comparing our *T. rex* not only to other *T. rex*es, but to discoveries, some of them just as new, of carnivores that came before *T. rex*, we can begin to figure out how *T. rex* came to be.

ABOVE: *T. REX*'S TOES WERE BIG AND EXTENDED THE SIZE OF ITS FOOT CONSIDERABLY. GREG PAUL AND BOB BAKKER SUGGEST THAT LONG TOES HELPED IN SWIMMING, BUT DUCKBILLS HAD SHORT, FLAT FEET, AND THEY WERE GOOD SWIMMERS. DEER SWIM WELL, AND THEY HAVE TINY TOES. *T. REX*'S TOES WERE POWERFUL, BUT IT'S HARD TO SAY IF THAT POWER WAS USED FOR SWIMMING, RUNNING, OR TEARING UP MEAT.

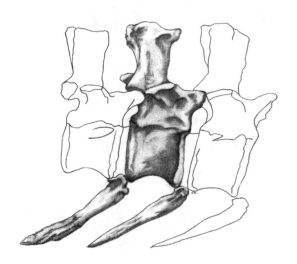

CHAPTER 6

T. REX AND FAMILY

TYRANNOSAURUS REX didn't just arrive on earth one day like an alien monster. It evolved from earlier animals, just as we have. Most people think of evolution, wrongly, as simply the triumph of the fittest, and *T. rex* certainly looks like a winner.

But like 99 percent of all the animals that ever lived, *T. rex* is extinct. That doesn't make *T. rex* a loser. Evolution is change, but it isn't necessarily progress. As wonderful and varied as life is, living things are limited in how much and in what ways they can change. And the pace of change can vary, from slow and steady to quick shifts and long stops. We have hints of this in the fossil record, of rapid evolution of some animals, stasis or slow change in others.

T. rex was one of the last and most spectacular products of dinosaur evolution. It was an experiment that can't be repeated, but it was no more a fluke or freak of nature than any other creature.

T. rex evolved within what is to us a tangled bundle of meat-eating dinosaur lineages, among animals whose origins go back hundreds of millions of years to some of the earliest dinosaurs known. We could of course trace *T. rex*'s ancestry back, very roughly, all the way to the dawn of animal life, in the oceans more than 600 million years ago. *T. rex*'s distant nondinosaurian ancestors first crawled up on land about 350 million years ago.

Instead, let's take a quick walk through dinosaur time.

T. rex's line, the order of dinosaurs, begins "only" another 125 million years after the first animals that came up on land. And dinosaurs were the dominant creatures for most of the 160 million years that followed their initial appearance.

We don't seem to realize that. Instead, we think of dinosaurs as huge things that died out, and call them failures. For longevity, they don't compare with horseshoe crabs or cockroaches, but dinosaurs did last a hundred times longer than we've been around, so far.

Nor do we stop to think what exactly it is that makes *T. rex* and its kin dinosaurs. It's not size. *T. rex* was horribly big, but most dinosaurs were smaller than bulls. And there were a lot of big animals that lived in dinosaur time that weren't dinosaurs. Other animals ruled the air—pterosaurs. And different animals, mosasaurs and plesiosaurs, were the giants of the sea.

Nor are dinosaurs necessarily dead. Many paleontologists think birds are the direct descendants of some meat-eating dinosaur cousins of *T. rex*, so you could easily argue that dinosaurs—small, flying ones—are alive today.

Some of the characteristics dinosaurs had in common they also shared with birds and other members of the groups we used to lump together as reptiles—scaly skin (look at a bird's legs) and young hatched from eggs. But what distinguishes dinosaurs from other reptiles is that they moved on land, most up on their toes with their legs beneath them, not splayed out. Their hips and ankles were constructed differently from those of other reptiles, allowing them to walk and run efficiently. There are other features of the skeleton that mark all dinosaurs, and some that separate one group of dinosaurs from another. But if an animal walked straight legged on land and had no fur, chances are it was a dinosaur.

The first dinosaurs evolved about 225 million years ago. The closest thing to a dinosaur ancestor that's been found yet is a house-cat-sized animal called *Lagosuchus* that lived in Argentina about 235 million years ago.

First animals

590m.

One of the oldest dinosaurs we know well also comes from Argentina. It wasn't a direct ancestor of *T. rex* and the rest of the dinosaurs, but it's the best we've got from that time. It's called *Herrerasaurus*, and it lived about 225 million years ago in the Triassic period (245 to 210 million years ago). Judging from a terrific skull University of Chicago paleontologist Paul Sereno found in Argentina in 1988, *Herrerasaurus* was a capable carnivore. It ran on two big hind legs and had huge teeth and a double-hinged jaw, just as *T. rex* did 160 million years later. But *Herrerasaurus* was more primitive in many ways, and a lot smaller than *T. rex*. It seems *Herrerasaurus* grew to "only" about ten feet long and five hundred pounds.

Over the next few million years two lines of dinosaurs arose—those with birdlike hips (ornithischians) and those with "lizard hips" or pelvises (saurischians). Confusingly, the latter, not ornithischians, are the group ancestral to birds. *T. rex* and the rest of the meat-eating dinosaurs, called theropods, and the giant "brontosaur" browsing dinosaurs, the sauropods, both belong to the so-called "lizard-hipped" line.

T. rex was one of the biggest of the theropods, but others were as small as chickens. All theropods were carnivores with many hollow bones. All walked on their back legs with three or fewer working toes on each foot. Some had flexible tails, while a more durable and equally varied group of theropods had stiff tails—the back

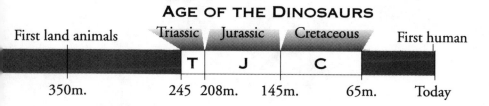

AGE OF THE DINOSAURS

First land animals | Triassic | Jurassic | Cretaceous | First human

T | J | C

350m. | 245 | 208m. | 145m. | 65m. | Today

bones at the rear end of their tails interlocked in various ways. These are called tetanurans, and *T. rex* is one of them.

We once split meat-eating dinosaurs into big theropods called carnosaurs and little ones called coelurosaurs. But since there was no real evolutionary connection between those in each group, we've stopped splitting them up that way. After all, you don't put elephants and giraffes into one group because both of them were huge. You do put elephants in the same group with little marmot-sized hyraxes because their skeletons show so many similarities that they must have evolved from a common ancestor. Lots of big dinosaurs had little cousins, too.

In the middle of dinosaur time, the Jurassic period (210 to 144 million years ago), the first tetanurans appeared. The most primitive we know of is the three-foot-long *Compsognathus*, one that walked over what is now the East Coast of the United States.

DINOSAURS FALL INTO TWO GROUPS ACCORDING TO THEIR HIP STRUCTURE.

TOP: AMONG BIRD-HIPPED (ORNITHISCHIAN) DINOSAURS, SUCH AS THE DUCKBILLS, THE PUBIC BONE IS TILTED BACK HORIZONTALLY, NEXT TO ANOTHER BONE OF THE PELVIS, THE ISCHIUM.

BOTTOM: IN LIZARD-HIPPED (SAURISCHIAN) DINOSAURS, SUCH AS *T. REX* AND OTHER THEROPODS AND THE GIANT BROWSING SAUROPODS, THE PUBIC BONE IS POINTED DOWNWARD AND ENDS IN A BOOT.

BARYONYX IS A RECENTLY DISCOVERED CARNIVORE FROM ENGLAND WITH A PECULIAR CROCODILE-LIKE SNOUT. IT MAY HAVE BEEN A FISH-EATER.

Late in the Cretaceous (the period from 144 to 65 million years ago), the scariest tetanurans of all came along, the tyrannosaurids. Where these giant predators came from isn't known. Maybe they came from the allosaur line of big predators, maybe from a common ancestor, along with the troodontids, a man-sized group of dinosaurs with many birdlike features. If you take all the features unique to tyrannosaurids, you can imagine a hypothetical ancestral tyrannosaurid. It would have been "small"—fifteen to eighteen feet long, with a long snout, long slender hind limbs, a pinched middle toe, and small forelimbs.

Most of us who work on dinosaurs think that the logical guess is that tyrannosaurs evolved from a more primitive meat-eating dinosaur. But which one we can't say yet.

Whatever it evolved from, *Tyrannosaurus rex* automatically by name belongs to that family of predators known as the tyrannosaurids, along with a few other kinds of meat eaters of similar anatomy. Each kind of

dinosaur we know represents a different genus. There are several genera (the plural of genus) of tyrannosaurs in the tyrannosaur family. Some we know far better than others. Though we have only eleven *T. rex* skeletons, *T. rex* is one of the best-known tyrannosaurs (see tyrannosaur chart).

Animals that are even more closely related than on the generic level we categorize as species. The usual definition of a species is a group of animals that look alike and can breed together. With living animals it's often easy to tell one species from another. The large cats belong to one genus (*Panthera*), but it's easy to see which species are lions and which are leopards.

You can't tell species so easily with fossil animals. You don't know anything about their interbreeding abilities, and often very little about their anatomy. Almost half of the six hundred or so species designations for dinosaurs are based on just a few teeth or bone scraps. So when we find something very reminiscent of another fossil we know already, but with differences we can't peg to a different sex or age, we create a new species for it. How closely related one fossil animal is to another is very much a matter of opinion. Some paleontologists

NOT ALL CARNIVOROUS DINOSAURS ARE INTIMIDATING. *COMPSOGNATHUS* WAS THE SIZE OF A ROASTING CHICKEN.

are "splitters." They see most every difference between skeletons as reflecting different species. Other paleontologists are "lumpers"—they tend to ascribe the differences between specimens to different sexes, ages, or other variations within a species.

Most of the time when we find an unfamiliar dinosaur, it looks so different from anything we've seen before that it rates a new genus (written in italics, first letter capitalized—e.g., *Tyrannosaurus*). We also give it a second or species name (written in lowercase italics—e.g., *rex*). We use Latin or Greek words a lot, by agreement, but you can name a dinosaur after anything. Many dinosaurs have been named for dead paleontologists and places where the fossils come from, but some have even been named for companies that lend money or equipment. With paleontologist Don Baird's advice, I named the first "feminine" dinosaur—the duckbill *Maiasaura* ("good mother reptile")—using the feminine sufix -*saura* instead of the masculine -*saurus* because the dinosaur was found with nests and babies. Tom Rich in Australia named the second feminine dinosaur, *Leaellynasaura*, for his daughter, Leaellyn.

Some of the dinosaurs that scientists have named as new species have turned out to be babies or juveniles of a species already named. All of us, scientists and dinosaur lovers alike, pay a lot more attention to the genus names than to the species names with dinosaurs. *T. rex* is the only dinosaur I know of that we commonly call by both its species and genus name, or by the first letter of its genus name.

But *Tyrannosaurus* is just one of several tyrannosaurid genera. So what do tyrannosaurids have in common? They are big and scary, but then so were many other dinosaurs. Tyrannosaurids were among the last dinosaur carnivores. They all lived toward the end of dinosaur time, during the Late Cretaceous period, 83 million to 65 million years ago.

As it turns out, all the tyrannosaurids had so much in common that Canadian paleontologist Phil Currie, who knows them as well as anyone, says they were as similar

PHIL CURRIE IS ONE OF THE FEW PEOPLE WHO CAN IDENTIFY ANY KNOWN CARNIVOROUS DINOSAUR FROM A SINGLE TOOTH. THIS TIME HE'S FOUND A CARNIVORE'S LEG BONE, IN THE GOBI DESERT OF CHINESE MONGOLIA.

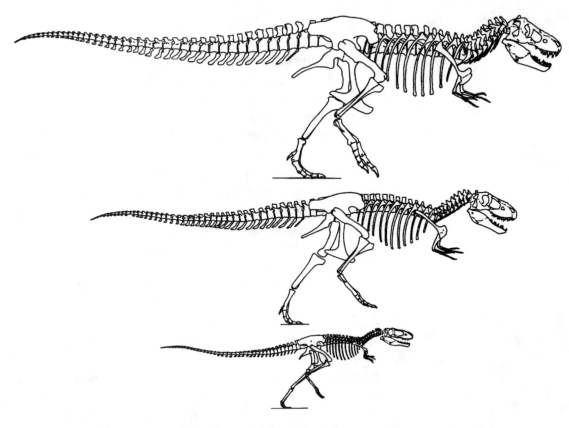

as "breeds of dogs, or models of cars." They shared innovations not seen before on big meat-eating dinosaurs—only two fingers on tiny front arms, and a heavily reinforced skull. Tyrannosaurids all had huge long teeth, curved and serrated on the edges. Their necks were short and thick, their chests broad and deep. Compared with tyrannosaurids' huge size, their tails were short and skinny, their hips narrow. Their legs were huge, with thick drumstick shins.

A lot of the other anatomical distinctions of tyrannosaurids may be too technical to be of much interest to anyone but us paleontologists, from the size and shape of backbones to the size and shape of teeth (all tyrannosaurid incisors are D-shaped when you cut across them) and skull openings in the cheek and eyebrow bones. These subtle distinctions are important because they show us how and how much the tyrannosaurid body plan changed from that of earlier

TYRANNOSAURUS REX (TOP LEFT) LOOKS IMPOSING EVEN WHEN COMPARED TO OTHER TYRANNOSAURIDS — *TARBOSAURUS, MALEEVOSAURUS, ALBERTOSAURUS* (TWO SPECIES) AND *DASPLETOSAURUS.*

carnivores. In contrast to earlier big meat-eaters, a
tyrannosaurid had narrower neck bones and skinnier
shoulder bones, but a much bigger skull and much
larger pubic bones.

The common ancestor of all the tyrannosaurs was
probably an animal a few million years older than
Albertosaurus, which lived about 75 million years ago.
George Olshevsky, a San Diego computer whiz, former
comic book collector, and chief keeper of the dinosaur
family tree, thinks the first tyrannosaurid may have
looked like *Alectrosaurus*, a questionable tyrannosaurid
that lived in Mongolia about 80 million years ago. We
don't know *Alectrosaurus* well, and its bones were
once mixed up with those of another dinosaur that had
large front limbs with big claws. But if you take away
those puzzling front limbs from *Alectrosaurus* (as paleontologists Bryn Mader and Robert Bradley did lately in
redescribing the animal) and fill in some missing parts

with guesswork, you can get an animal close to that make-believe tyrannosaurid ancestor I described earlier—a svelte, eighteen-foot-long hunter with a long low head and long skinny legs. It isn't such a big evolutionary step from this recreated *Alectrosaurus* to *Albertosaurus*, and on to a variety of other tyrannosaurids, ending in *Tyrannosaurus rex*.

Which was *T. rex*'s closest relative? Probably

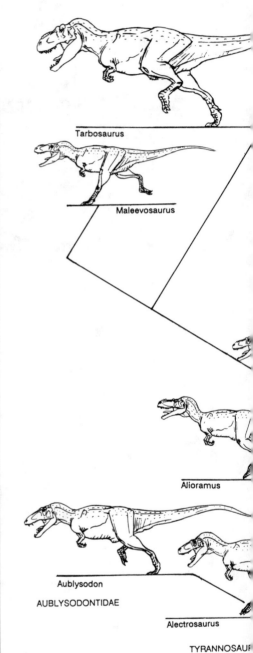

Tarbosaurus

Maleevosaurus

Alioramus

Aublysodon

AUBLYSODONTIDAE

Alectrosaurus

TYRANNOSAUR

THE TYRANNOSAURID FAMILY

*I*t is by no means clear which animals truly belong to the tyrannosaurids. Several scientists are reviewing the fossil evidence now, and new finds are made all the time, so chances are this list, like all others, will soon be altered. For now, George Olshevsky lists the following genera and species of tyrannosaurids:

Albertosaurus megragracilis was named by artist Gregory Paul in 1988. This was a smaller, more lightly built species of albertosaur, still as big as a rhinoceros.

Albertosaurus sarcophagus was named by Henry Fairfield Osborn in 1905. It was smaller than *T. rex*, with a lower, longer snout and grew to twenty-six feet and two tons.

Daspletosaurus torosus was named by Dale Russell of Canada in 1970. A solidly built predator with a huge head, legs, and tail, it grew to twenty-eight feet and three tons.

A new "stretch-snouted" *Daspletosaurus* will be named by Bob Bakker.

Gorgosaurus libratus was named by Lawrence Lambe of Canada in 1914. It might have been a large *Albertosaurus*.

Gorgosaurus sternbergi was named by American Museum paleontologists William Diller Matthew

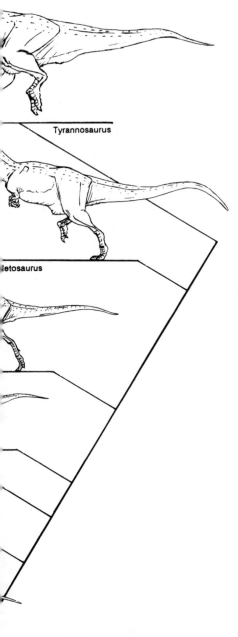

Tyrannosaurus

letosaurus

Tyrannosaurs

...NOSAURIDAE

and Barnum Brown in 1923. Relatively small compared with *Gorgosaurus libratus*, it might be a different genus of tyrannosaur entirely.

Nanotyrannus lancensis was named by Bob Bakker, Mike Williams, and Phil Currie in 1988 from a Cleveland Museum skull originally identified as *Gorgosaurus lancensis* by Charles Gilmore in 1946. This "pygmy" tyrannosaur had a long snout and grew to "only" fifteen feet long.

Tarbosaurus bataar was named by Evgeny Maleev of the USSR in 1955 from a find in Mongolia. It grew up to thirty-three feet long and resembled *T. rex* closely.

Tyrannosaurus rex was named by Henry Fairfield Osborn in 1905 from Barnum Brown's discoveries in Montana. Bob Bakker thinks *T. rex* might be two species.

LIKELY TYRANNOSAURIDS

Alectrosaurus olseni was named by Charles Gilmore in 1933 from bones found in the Gobi Desert of Inner Mongolia. Gilmore may have combined two different animals. Ken Carpenter thinks *Alectrosaurus* is not a tyrannosaurid, but Phil Currie believes it was a primitive tyrannosaurid. It was about sixteen feet long and less than half a ton in weight, with skinny shoulders and legs.

Alioramus remotus was named by Sergei Kurzanov of the USSR in 1976 from a skull and a few bones from Mongolia. *Alioramus* was a mysterious creature about twenty feet long. The back of the skull looks like a tyrannosaurid, but *Alioramus* had no holes in its cheek and no eyebrow bones like other tyrannosaurids. It also had lots of little horns on its muzzle, something not seen in any other tyrannosaurid. Ken Carpenter excluded *Aliosaurus* from tyrannosaurids in his reclassification of the family. It may be related to *Alectrosaurus*. Once again, Phil Currie says it definitely was a tyrannosaurid, most closely related, by braincase comparison, to *Tarbosaurus*.

Chingkankousaurus fragilis was named by C. C. Yang in China in 1958. A big theropod, it is known only from the shoulder blade and other fragments and may be a *Tarbosaurus*.

MAYBE TYRANNOSAURS, SOMEDAY

A new genus of tyrannosaurid, or at least a new species of *Daspletosaurus*, may be described soon from the Horseshoe Canyon Formation of Alberta, Canada, by Bob Bakker, Mike Williams, and Phil Currie.

A new species of tyrannosaurid may be described by Bakker and Currie from the tyrannosaurid on display at the Field Museum in Chicago now labeled *Albertosaurus libratus*.

A new species of tyrannosaurid, an intermediate between *Daspletosaurus* and *Tyrannosaurus rex*, will be described by Dave Varricchio from a complete skull with articulated leg found at a Museum of the Rockies' site in western Montana.

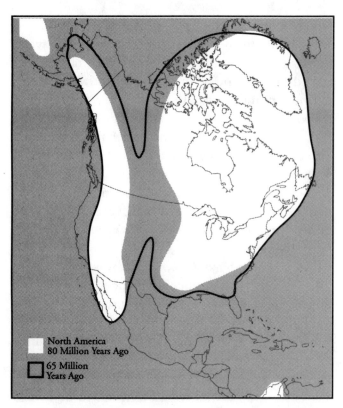

North America
80 Million Years Ago

65 Million
Years Ago

T. REX COULDN'T HAVE VISITED NEW YORK IF IT HAD WANTED TO. A SHIFTING INLAND SEA SEPARATED EASTERN AND WESTERN NORTH AMERICA THROUGHOUT MUCH OF LATE CRETACEOUS TIME, INCLUDING THE REIGN OF *T. REX*.

Tarbosaurus bataar, a big predator with a striking resemblance to *T. rex*. Skeletons of *T. bataar* have been found several times in Mongolia's Gobi Desert, the first immediately after World War II. A Russian paleontologist suggested that *Tarbosaurus* was intermediate between *Albertosaurus* and *Tyrannosaurus*, but the truth is *Tarbosaurus* is a lot closer to *Tyrannosaurus* than anything else. You have to look very hard to find the differences between *Tarbosaurus bataar* and *Tyannosaurus rex*, so hard that *Tarbosaurus* is considered by some an invalid genus. Instead, the Mongolian giants could be grouped as *Tyrannosaurus bataar*, in the same genus as their American cousin, *T. rex*.

Ken Carpenter, who did that regrouping, gives a telling example of how minor the differences are between *T. bataar* and *T. rex*. The back of the upper jawbone in *T. rex* ends before the front rim of the eye socket. In *Tarbosaurus*, the bone ends farther back, under the front part of the eye socket. For another, *T.*

bataar's tiny arms are even smaller than *T. rex*'s.

Ken thinks the minor nature of differences between *T. rex* and *Tarbosaurus* suggests they are closely related species. And he thinks *Tarbosaurus*'s features are more primitive than those of *T. rex*. So he thinks *Tarbosaurus* was *T. rex*'s ancestor. Perhaps some *Tarbosaurus* crossed over the link that had emerged between Asia and North America and developed into *T. rex* in this continent at the very end of the Cretaceous period. (A few teeth and jaws found in China suggest *Tarbosaurus* lived there also, though the Chinese have named three other species of *Tyrannosaurus* from these fragments.)

Not everyone thinks *Tarbosaurus* was *T. rex*'s older sister species. Some other differences between the animals are more substantial—for instance, *T. bataar* was more lightly built. Now some studies are underway that suggest those differences are big enough after all for the Mongolian tyrannosaurs to rate their own genus. Then they would have been older cousins of *T. rex*. In that case *Tarbosaurus* would turn out to be a valid genus after all.

These name changes are confusing for all of us. But that's how it often goes with classifying dinosaurs. We know so little about them that when we take a closer look or get a bit of new evidence, the categories we've squeezed them into don't seem to fit anymore.

Another member of the tyrannosaurid family has only recently been identified, though it sat gathering dust in a museum basement for decades. Inspired by *T. rex* discoverer Barnum Brown, paleontologists from the Cleveland Museum of Natural History went to *T. rex* country to get a dinosaur in 1942. They found a skull of a small tyrannosaurid. In 1946 Charles Whitney Gilmore of the Smithsonian identified it as a new kind of gorgosaur, *Gorgosaurus lancensis*. In 1970, Canadian dinosaur paleontologist Dale Russell decided it was a dwarf species of *Albertosaurus*.

Paleontologist Bob Bakker came across the skull on a visit to the Cleveland Museum in 1988. Bob and paleontologist Mike Williams of the museum had their doubts about whether a skull that small, narrow, and otherwise peculiar could belong to *Gorgosaurus*. The more they examined the skull, the more they became convinced that this was not a juvenile *Gorgosaurus*, but an adult of something else. Their suspicions were confirmed when Mike Williams chipped away the horns of the skull and found they were pure plaster. An overeager fossil worker had added a couple of plaster horns to the skull to make it look more like the *Gorgosaurus* known elsewhere.

Bob, Mike Williams, and Phil Currie renamed the animal *Nanotyrannus* ("little tyrant"). *Nanotyrannus* had things in common with *T. rex* and no other dinosaur. The skull had a narrow, doglike face—it was far wider in the back than across the snout. It had no horns over its eyes, and the eye sockets were pointed so far forward that they may have seen with binocular vision.

T. rex and *Nanotyrannus* lived at the same time. They're found in the same rock formation. If *T. rex* and *Nanotyrannus* are not from the same genus, then they may have evolved their matching characteristics independently from entirely different stock. Bob Bakker thinks the latter—that they evolved from very different ancestors.

CAT SCANS OF THE SKULL OF
NANOTYRANNUS SHOW THE
DELICATELY SPIRALING
TURBINAL BONES
OF THE NOSE.

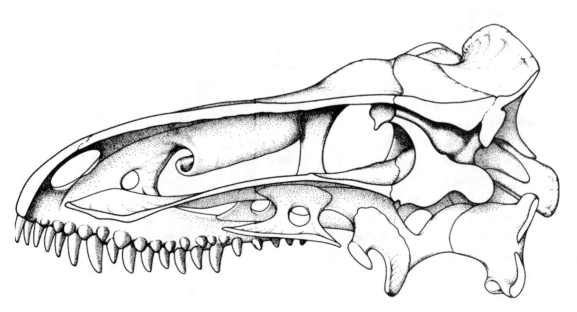

Other dinosaurs, much less completely known have been pigeonholed into the tyrannosaur family (see sidebar). We may yet find more of these animals, enough to tell whether they were something new or something familiar after all. Certainly there are a lot more carnivorous dinosaurs out there. As they are found, they will answer more of our questions about how many tyrannosaurs there were, where they lived, and how they evolved.

We may not have to wait all that long for some answers. We dug up a beauty of a juvenile *Allosaurus* in Wyoming in 1991. Bones of young allosaurs have been found before—by the truckload at the Cleveland-Lloyd quarry near Price, Utah—but never before in a single skeleton so well preserved. This one should tell us a lot about how *Allosaurus* looked as it was growing up.

Phil Currie and his colleagues from the Royal Tyrrell Museum of Palaeontology in Alberta have found a half-dozen *Albertosaurus* skeletons in the decade since the museum opened. Another *Albertosaurus* was recently found in Alabama. That discovery shows that tyrannosaurids lived east of the big seaway that ran north-south through the heart of North America at the end of dinosaur time.

And I've been finding lots of interesting dinosaurs, including big carnivores, throughout the Two Medicine Formation of western Montana. The Two Medicine Formation is more than two thousand feet thick there, full of fossils from 6 million years of late dinosaur times. And we've been able to link particular kinds of dinosaurs to the shifting pattern of the inland sea that stretched across Montana at that time. As the seaway grew, or trangressed, dinosaurs (and other land animals) were pushed up into isolated upland habitats, such as mountain valleys. When the seaway shrank, dinosaurs flourished in delta environments. (I describe this geology more in my book *Digging Dinosaurs*.)

Not only did the kinds of dinosaurs vary with the changing habitats, but the way in which they evolved changed according to the environment. It was a stressful time when the seaway expanded, encroaching on the dinosaurs' turf, and as you might expect, it was the less specialized species that survived in those conditions. They were more adaptable to changing environments. There was less diversity and smaller populations of dinosaurs when the seaway moved in. But new species still evolved. It was straight-line evolution, however, one species evolving into just one other, instead of radiating into several. We call this straight-line evolution anagenesis.

When the North American seaway closed in on dinosaur habitats in the Two Medicine Formation of Montana where I've been exploring, new species evolved, in a straight-line path. Sometimes they just got bigger. But one kind of duckbill led to another, one horned dinosaur to another.

When the seaway receded, new territories opened up. Dinosaurs radiated into new species to meet the opportunities to fill these new niches, an example of cladogenesis. One population of animals enters into an environment with many niches and evolves

DASPLETOSAURUS WAS A CLOSE RELATIVE AND POSSIBLE ANCESTOR OF *T. REX*.

into many species to fill them.

We see these two patterns of change, straight-line and branching, primarily with the dinosaurs we have the greatest samples of, duckbills and horned dinosaurs. Right after the end of a major transgression, or inland movement of the sea, ends, we see the greatest diversity in dinosaurs. In the last few years we've collected four thousand specimens from these few million years in dinosaur time in one area. One hundred of those dinosaurs can be identified to the species level. That's more of a record than anyone's ever had for dinosaurs. This is a dramatic confirmation of the theory of evolution itself. To me it's about the most important finding of all my work on dinosaurs.

What does all this have to do with *T. rex*'s evolution? Well, it was during that stressful time recorded in the top of the Two Medicine that *Daspletosaurus* was replaced by a tyrannosaur forerunner of *Tyrannosaurus rex*. We know that because we've found a predator intermediate in appearance between those two animals, with no features peculiar to it alone. That shows anagenesis was going on in tyrannosaurid evolution at that time. Intermediates, like this new tyrannosaurid,

COELOPHYSIS WAS ONE OF THE EARLIEST CARNIVOROUS DINOSAURS KNOWN. NOT MUCH BIGGER THAN I, IT LIVED IN WHAT IS NOW ARIZONA IN THE FIRST DINOSAUR PERIOD, THE TRIASSIC, SOME **220** MILLION YEARS AGO.

in straight-line evolution between animals, are rare. Before I wrote about the intermediates (metataxa) we found among duckbills, horned dinosaurs, and tryrannosaurids, there were only two dinosaur metataxa known—*Coelophysis*, an early predator, and *Archaeopteryx*, the earliest bird.

From this new discovery it now looks as though *T. rex* wasn't a particularly sturdy branch on the spreading bush of big predators. More likely it was the one descendant of a big theropod, *Daspletosaurus*, nearly as big as *T. rex* itself. At least that's what the little evidence we've got now says. We'll know even more about the relationship between the two big predators when my doctoral student Dave Varricchio finishes his thesis research on *Daspletosaurus*.

But there are still many mysteries to be solved about *T. rex*'s evolution. I think it's only a matter of time before we find more tyrannosaurids, enough to tell us more about where *T. rex* came from. But unlike the other tyrannosaurids, *T. rex* didn't lead anywhere—it never evolved into anything else. There never was another tyrannosaurid after *T. rex*. I think that's a shame.

ARCHAEOPTERYX IS ONE OF THE FEW KNOWN METATAXA, A TRUE INTERMEDIATE ANIMAL, IN THIS CASE BETWEEN BIRDS AND DINOSAURS.

CHAPTER 7

THE WORLD OF T. REX

WHERE DID *T. REX* LIVE?
Anywhere it wanted to, as the thousand-pound-gorilla
joke goes. I'm not being a wise guy.

Whether *T. rex* was a scavenger or a predator (see
Chapter 9), I can't imagine what kind of habitat it might
have liked best, though most modern animals have a
preference for one type of environment. It wouldn't
have mattered what kind of plants it was around. It just
needed to be around some meat. And I don't think it
cared whether the meat was from the lowlands or the
uplands.

Just what the weather was and what the environment
looked like 65 million years ago isn't certain. The best
evidence we have for answering those questions comes
from fossilized pollen, leaves, and trees. Tree rings can
show seasonal growth and cold or dry winters. The sizes
and shapes of fossil leaves tell about temperature and
humidity. If leaves were large and green year-round with
pointy drip-tips, then the climate was warm and wet. If
they were small and rounded, the weather was drier or
colder. Deciduous tree leaves indicate more seasonal
temperature, moisture, or light.

THE PLANT EVIDENCE

Finding fossil plant evidence isn't easy. Fossil bones are
usually preserved in stream channels. Fossil plants turn
up in stream channels and also in floodplains, ponds,
lakes, and deltas. In ancient stream channels, at least,
only the bigger, sturdier leaves, plants, and trees become fossilized. They're the only ones strong enough to

hold up to the force of the water that was moving them and the mud that covered them at some break in the flow where debris piled up.

But where you find fossil bones you don't usually find plants. The acid environments that preserve leaves break down bone. So when fossil leaf-hunters go to the Hell Creek Formation, they head for the brown bands in the cliffs, rock that's turned dark with the carbon from decayed plants. Often it's a layer of clay or shale right above a black coal stripe that doesn't preserve anything you can see with the naked eye. The brown and black layers, however, are just where dinosaur diggers don't go to look for bones. We find bones in the tan sandstone and gray mudstone. We find plants in the parts of the Hell Creek Formation closest to the ancient shoreline, in what are now the Dakotas. Those waterlogged sites protected plants from drying out and breaking down in air.

There are exceptions to every rule. If you look long enough and hard enough, you can find plant fossils with dinosaur bones. Leo Hickey, a Yale University paleobotanist, found plant fossils underneath a *Triceratops* near Jordan, and his former student Kirk Johnson (now of the Denver Museum of Natural History) finds plant fossils with fossil bones in North Dakota.

We're sifting through bags and bags of rock matrix from around our *T. rex,* hoping to find plants. We can't see much in the way of plants from our quarry so far, but that's how it usually is. If we find plant remains with our *T. rex,* they're more likely to be pollen than leaves. (Fossil trees are very hard to come by—though a 165-million-year-old fossil forest was recently found in northwestern China.) And it isn't easy to know when you've got hold of pollen. Each pollen grain is as small as a dust particle—you need to look at them under a microscope at $100\times$ to $800\times$ magnification to see them clearly.

Tiny as they are, pollen grains are amazingly durable. Each grain is coated with a waxy surface that will preserve it pretty much forever if it is in an acidic environment, not exposed to oxygen. We've got pollen grains from long before the age of dinosaurs, going back more than 360 million years. To extract the pollen you can pour acid onto the rock in which pollen is found, then spin it in a centrifuge. Once the rock has dissolved, pollen sinks to the bottom. You spread that slurry on a slide, put it under the microscope, and look for pollen grains. Different kinds of plants produce different-looking pollen grains. Some, such as grains from pine trees, have bladders that look like water wings. Others have spikes like a medieval mace.

Fossil pollen experts (called palynologists) have found well over one hundred kinds of pollen grains in the Hell Creek Formation. Pollen is a pretty powerful tool for dating rocks and the fossils in them. Pollen is so common and so durable that you can find millions of pollen grains in a single ounce of rock. Still, there is only so much you can figure out about the environment from pollen. The fossil pollen looks something like that from plants today. We can tell some of it came from conifers, some from sycamores, and some from ferns, palms, and extinct broadleaf trees.

Fossil leaves tell us more. Despite the obstacles to leaf preservation, fossil plant experts have found many sites in the Hell Creek Formation that have plant fossils in them. Leo Hickey and Kirk Johnson have dug hundreds of holes in these badlands looking for leaves. They've come up with more than a hundred good plant fossil localities, the overwhelming majority of them in the Dakotas. In all, Kirk and Leo have examined nearly twenty thousand leaves from the Hell Creek Formation.

Kirk has studied leaves found at the sites of the two Black Hills *T. rex* discoveries in South Dakota. From the "Stan" site Kirk saw mostly flowering plants, some of them relatives of magnolias, sycamores, and the laurel family (which includes avocados). Less than 10 percent of the leaves were conifer needles. They included a relative of the sequoia and another more surprising conifer, the monkey-puzzle tree. That group of conifers is found now only in the Southern Hemisphere, but in dinosaur times they grew worldwide. A Jurassic tree of

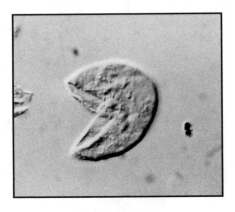

THIS PAC-MAN DESIGN BELONGS TO THE EVERGREEN POLLEN GROUP, *TAXODIACEAE POLLENITES HIATUS*, SUGGESTING IT IS RELATED TO *METASEQUOIA* AND THE BALD CYPRESS.

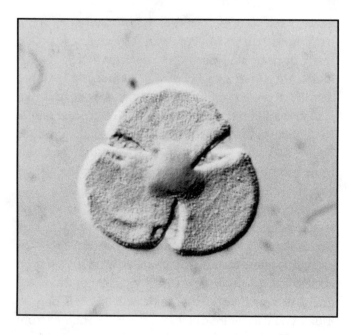

this family is the biggest tree ever known—its partial trunk is a fossil 380 feet long. The tree itself was probably 460 feet high, nearly 100 feet taller than the tallest tree today.

Kirk also studied leaves found in the body cavity of "Sue." These were a different mix of plants, unlike any Kirk had seen in a hundred Hell Creek sites. Types of plant that were not peculiar, just the relative abundance of some kinds. For instance, he found fewer flowering plants and lots more sequoia-like conifers, ferns, and aquatic plants than he expected.

There are two good reasons why what we know of Stan's world looks so different from Sue's. We're not getting a complete picture of either, only a sample of what happened to be preserved. Sue was buried in a stream channel, and many plants from distant places may have washed in among the bones. Stan comes from a point bar, a higher elevation more likely to preserve what actually grew nearby. And Stan and Sue came from very different times. Stan comes from very near the end of Hell Creek Formation time, sediments close to the boundary at the end of the Cretaceous. Sue comes from near the middle of Hell Creek time. That's a differ-

THE THREE-LOBED WHEEL IS *GUNNERA MICRORETICULATA,* ONE OF THE FEW FOSSIL POLLEN GRAINS FROM THE HELL CREEK FORMATION THAT HAS A KNOWN RELATIVE TODAY. ITS MODERN COUSIN IS A LARGE-LEAVED FLOWERING PLANT OF THE MOIST TROPICS.

ence of millions of years, and the climate and vegetation changed a lot in that time. Until the last few million years of dinosaur time in the Hell Creek Formation, plant samples show a mix—half ferns, the rest pretty evenly divided between flowering plants and conifers. In the last 2 million years of the Hell Creek Formation, the plant samples show almost all flowering plants. As Kirk puts it, "that's a fundamental reorganization of vegetation, a bigger shift than at the boundary" of dinosaur time.

We're only beginning to fully understand those changes in plant life. People have been looking for and finding *T. rex*-aged leaf fossils for decades, but a lot of the time they misidentified the leaves they found. That's why if you go into the Late Dinosaur Hall of some of the grand old natural history museums and look at the mural, you'll see oaks, elms, and willows in the background, and maybe even some grass in the foreground. We know now those plants didn't exist in *T. rex*'s time. Most of what we find are extinct kinds of plants. They might belong to a living family of plants, but they were different from anything we know today. What we can say with certainty is that *T. rex*'s world was full of many kinds of plants, far more kinds of trees than we'd find in western North America today.

We know more than two hundred kinds of plants from *T. rex* country, more than 90 percent of them flowering plants. Since only selected parts of certain habitats get preserved as fossils, those two hundred plants are only a fraction of what must have grown in Hell Creek Formation times. The actual diversity of plants was probably closer to what you'd find today in a rainforest as opposed to any temperate habitat.

Just what communities those plants lived in is hard to say from clumps of leaves washed together from various delta locales. We'd have a better answer if we could find a piece of dinosaur land preserved in place just as it was.

That's a lot to wish for, but if you're lucky, you might find such a place. In the summer of 1990, Scott Wing, a Smithsonian paleobotanist, got lucky. He was looking at a rock outcrop near the town of Worland, Wyoming. The rock is from the Meeteetse Formation, and it's 72 million years old, close to 7 million years older than the Hell Creek Formation, from which we dug up our *T. rex*.

Scott Wing came upon some volcanic ash (bentonite), and just beneath it a patch of open vegetation exactly as it looked in dinosaur days. He'd dug up a sort of Pompeii for plants: fossils created when mounds of cool ash from a nearby erupting volcano were washed over the ground. The area Scott found preserved by the volcanic eruption stretches more than a mile by two miles. In it, Scott found more than 110 species of plants, sometimes preserved nearly whole.

From one single pit three feet by six feet and a foot and a half deep, Scott and Leo Hickey dug up more than fifty different kinds of plants. You don't find those numbers in many places in the world today. Many of the plants they found are delicate little herbs that would never have been fossilized in the usual stream channel fossil sites. Others were big palm fronds.

This sort of concentration of plants is extremely rare from late in dinosaur time. But those were sites that missed out on low ground plants. Low-growing herbs made up less than 15 percent of previous samples of plants from that time. At Scott's site, however, these small broadleaf flowering plants made up more than 60 percent of the kinds of species, but just 11 percent in relative abundance of what's found (the remainder were ferns, cycads, and conifers, including a cone from a monkey-puzzle tree). As a result, our impression of what some landscapes looked like near to *T. rex*'s time has now been greatly changed.

At another site sixty miles closer to the ancient volcano, Leo found standing tree stumps that were partly petrified, with the rest burned by hot ash. The environment Leo discovered seems to have been a forest of large conifers with ferns as a heavy ground cover.

What Scott Wing and Leo Hickey's discoveries provide us are our first snapshots of the vegetation on the

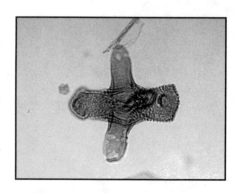

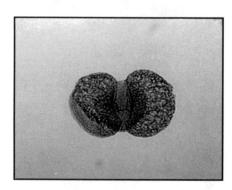

THE MOST ABUNDANT OF FOSSILS IS POLLEN. MICROSCOPIC GRAINS LIKE THESE FROM THE HELL CREEK FORMATION ARE FOUND BY THE MILLIONS.

ground close to *T. rex*'s time and neighborhood.

We know *T. rex* itself from places that in its time, 67 to 65 million years ago, were part of a huge web of floodplains. This delta occupied nearly one million square miles of the North American west. It lay east of the newly rising Rocky Mountains and their active volcanoes, catching the dirt and debris that washed down from the highlands. Large rivers emptied into the shallow and shrinking north-south seaway that then severed North America, from the Arctic to the Gulf of Mexico. The flow of water also formed lakes and low ridges along stream channels in a flat and largely featureless coastal plain.

We get dinosaur fossils only from deserts or places like these floodplains, where sediment builds up over bones (see Chapter 1). So we don't know whether *T. rex* lived in the eroding highlands closer to the Rockies. The two *T. rex*es found in Canada were from slightly more upland environments than the *T. rex*es from the floodplains of eastern Montana, Wyoming, Colorado, and South Dakota.

The fossil plant evidence we have suggests that the terrain looked nothing like the badlands of today. *T. rex* country "looked a lot more like New England in late spring today than like most other places," says Kirk Johnson. The land would have been green and lush, at least in the floodplains, with fern glades and a ground cover of small flowering herbs and creeping vines. Around the stream margins were little water lily–like plants. Small trees, palms, and ferns were prevalent.

T. rex lived 2 million to 7 million years after the time of the dinosaur environment Scott Wing found buried beneath a volcano's dust. But the environment in *T. rex*'s day was probably similar in a lot of ways. Maybe it was even more lush.

Paleobotanist Kirk Johnson and artists are now busy creating the first diorama of a Cretaceous forest ever illustrated from fossils, for the Denver Museum of Natural History. They're drawing the plants in the Hell Creek Formation, using a camera lucida (a mirrored

DINO DAMAGE: THE EFFECT OF DINOSAURS ON THEIR WORLD

*W*ith all these trees around, there was plenty for a foraging dinosaur to eat, whether nipping at seeds and fruits from tree branches, snapping up the ground cover in the forest, or wading into ponds and streams to eat the plants that grew in the water. What effect did *T. rex* and other dinosaurs have on *T. rex*'s environment?

T. rex probably didn't have a taste for plants. But other dinosaurs did, and we have a pretty good idea what plant-eating dinosaurs ate. It's not just indirect evidence, like the fact that they had flexible jaws and the best grinding teeth ever. We have hard proof—dinosaur poop.

Fossilized dinosaur scat is called coprolite. A good coprolite isn't hard to recognize. It has the same cylindrical shape as what we and Rover and many other animals turn out. And it isn't much bigger. But dinosaur poop is dry, hard, heavy, and germ-free. That's because it's been fossilized. But that doesn't mean all the organic parts are gone.

A former student of mine, Karen Chin, now doing her graduate work at the University of California, Santa Barbara, specializes in dinosaur coprolites. She has a good sense of humor about it, but don't ask her how her work is coming out.

Karen made a detailed study of what she calls a "top-of-the-line" coprolite my assistant Bob Harmon found near some plant-eating dinosaur bones around Cut Bank in western Montana. She made thin sections of the coprolite, then examined its components with X rays and analyzed them with a gas chromatograph. Karen found chewed pieces of leaves, stems, seeds, and resins from conifers, as well as algae and bacteria. She found waxes that indicated the animal was consuming the more advanced vascular plants whose leaves pro-

A SIERRA FOREST WITH SEQUOIA AND FIR TREES.

duce those waxes.

Karen's finding confirms what all of us would suspect—that plant-eating dinosaurs munched on the trees and plants around them. It makes sense that dinosaurs would be keeping the vegetation down. They must have had enormous appetites. Consider the great impact elephants have upon the African landscape. Plant-eating dinosaurs must have done at least as much to crop the vegetation around them. Maybe the openness of Scott Wing's site reflects the effects of dinosaur grazing, although the absence of plants could just as easily have been the result of fire or flooding. Closed forests appear to have been more extensive at that time.

In fact, dinosaurs do not appear to have had a drastic effect on these environments. As Leo points out, dinosaurs would have been pruning the branches of large trees and lopping the tops off saplings. But the trees appear to have thrived anyway. Maybe the trees had adapted to withstand dinosaur pruning. Relatives of these conifers today reach 120 feet to 180 feet in height and hold most of their nutrition in the foliage and small branches. They self-prune all of their lower lateral branches once they grow above eighty feet high and develop an umbrella-like canopy of branches, Leo says, "almost as if they retain a genetic 'memory' of the pruning they must have taken from the dinosaurs—and so are trying to cut their losses."

Bob Bakker suggests that dinosaurs altered the entire course of plant evolution, with the advent of efficient duckbill browsers favoring the development of flowering plants some 100 million years ago. To Leo Hickey this is "hogwash." The effects of fires, floods, and shifts in stream channels created far greater disturbances. Plants evolve very slowly compared with animals, and there's no evidence that dinosaurs directed plant evolution. If anything, I think it was the other way around. But I doubt we'll ever have any way of knowing that.

device invented in 1807 and still used) to project magnified images of the fossils themselves onto a flat surface for tracing. The display will include conifer trees of a variety of species that also grew in dinosaur habitats, particularly in the uplands. Some of these were relatives of junipers and may have looked a lot like them.

T. rex's country was, as far as we can tell, mostly a closed, high-canopy woodland. Some herbs, ferns, and perhaps pachysandra covered the areas of open land, but there were fewer of these plants than in earlier times. In moister areas, such as ponds and backwaters, the tallest trees were relatives of the bald cypress of today's southern swamps, but with leaves that didn't narrow to a little stem. Most of the trees were big-leafed kinds, like those of the laurel family, that kept their leaves year-round. Some were extinct relatives of modern conifers like the incense cedar, with scaly foliage, though at the very end of dinosaur times climatic changes seem to have killed off many of the conifers. Sticking out above the canopy were some giant redwoods and a few *Metasequoia*s.

The dawn redwood, the eerie *Metasequoia*, is one of my favorite trees, and it's among those that dwarfed even the biggest dinosaurs. *Metasequoia*s are strange, beautiful trees. They look something like bald cypresses, only they don't live by the water, and something like redwoods, but they have softer needles (and unlike redwoods, *Metasequoia*s shed their needles in winter). They're several feet in diameter when mature, all billowy and cone topped. They throw out shoots from old wood in their trunks, and they're incredibly knobby, with foliage everywhere. There was a full and bushy one with delicate little leaves at Princeton that I loved to sit under and imagine that I was in dinosaur time.

We know what *Metasequoia*s looked like through a lucky accident. *Metasequoia*s were known only as fossils until World War II, when a forester found an isolated grove still alive in China. The tree has now been cultivated and preserved. There's a nice one at the Harvard arboretum in Boston. You can walk underneath and imagine what it would be like to see the same tree from a dinosaur's eye.

T. REX WEATHER

Knowing the plants tells us something about the climate in *T. rex*'s time. We have modern relatives for many of these plants and so we know what climates suit them best.

According to these analogies, during most of *T. rex*'s time, the climate was what you'd expect to find on average in North Carolina, but without so much change between seasons. Light-sweater weather. But from shifts in the types of plant fossils, paleobotanists Kirk Johnson and Leo Hickey figure the climate changed considerably during *T. rex*'s time, the 2.5 million years that are recorded in the Hell Creek Formation.

KIRK JOHNSON COLLECTED THIS BEAUTIFUL FOSSIL LEAF IN MONTANA OF AN EXTINCT MEMBER OF THE LAURALES, A GROUP THAT CONTAINS AVOCADO AND CINNAMON TREES.

RIGHT: MARJORIE LEGGITT OF THE DENVER MUSEUM OF NATURAL HISTORY DREW THIS RECONSTRUCTION OF THE LAURALE LEAF AS IT LOOKED ON A TREE IN *T. REX*'S ENVIRONMENT IN MONTANA, 65 MILLION YEARS AGO.

Overall, the Hell Creek Formation preserves a time of warming with lots of new plants moving in. The weather was in the low fifties on average, as it is in the redwood forest, though not so wet through most of Hell Creek time. We certainly don't think the weather in *T. rex*'s day was dry—we don't find any caliches, the little balls of calcium carbonate that form in sediments from years of dry seasons. We do see those caliches in the pits we dig for duckbilled dinosaurs from more than 70 million years ago in western Montana. That's closer to the age of Scott Wing and Leo Hickey's plant fossils but near to the latitude of our *T. rex* find. So we conclude that the weather was getting moister by *T. rex*'s time.

But in the middle of *T. rex*'s time, according to Kirk, a relatively short drought happened, maybe lasting just tens of thousands of years. Perhaps the mean annual temperature dropped a few degrees as well. The proof for that is a band of rock with unusually small leaves in it, some of them relatives of modern roses. Without much water, leaves don't get big. And small-leafed plants need less water than big-leafed plants.

For the last nearly million years of dinosaur time, years marked by the uppermost stripes of the Hell Creek badlands, the plants were warm-weather types. Their fossils resemble those you'd see in older rocks from places nearly eight hundred miles farther south, such as New Mexico and southern Colorado. The average temperature rose by about a dozen degrees, making the Hell Creek into a lush subtropical land with warm weather year-round.

Still there had to be seasons in *T. rex*'s time. Montana was a little farther north than it is today (about fifty degrees north latitude), so in summer the days were considerably longer and brighter than in winter. At times the floodplains would have been downright hot and soggy, especially in swampy areas near the inland sea. Sudden storms produced flash floods that killed dinosaurs, buried them, and eventually turned some to fossils. Upland and away from the water, temperatures

THE FOLLOWING PICTURES ARE BY MONTANA PAINTER DOUG HENDERSON WHO HAS PHOTOGRAPHED CALIFORNIA'S SIERRA NEVADA, A MODERN ENVIRONMENT WITH MANY SIMILARITIES TO *T. REX*'S.

would have been more extreme. There fewer plants, and animals, including *T. rex*, could make a living.

Somehow *T. rex* and and the rest of the dinosaurs stopped making a living, everywhere, about 65 million years ago. Maybe the weather was slowly getting more extreme as the age of dinosaurs neared its end. I'm not big on extinction theorizing (see Chapter 1), but I think gradual, significant degrading of the climate over a longer stretch of time, maybe millions of years, would have been enough to do in the dinosaurs.

That's what Leo Hickey thought, too, up until about ten years ago. But fossils he and Kirk Johnson have found more recently made him change his mind. The evidence for a sudden change in the plant world at the Cretaceous boundary is a lot stronger than it is for dinosaurs. You can find certain pollen within one millimeter below the boundary (and none of that kind of pollen above the boundary), but there are no dinosaurs within several feet of that boundary. The weather did change significantly at the boundary, 65 million years ago. It got a lot warmer, up to a mean of sixty-one degrees Fahrenheit year-round in Hell Creek localities— about what it's like in São Paolo, Brazil, today. Then, after the dinosaur extinction, the weather got cooler again, dropping back to fifty-two degrees Fahrenheit within 2 million years. The way Leo sees it, the dinosaurs were running along happily right until the big chill. And the asteroid impact probably triggered the dip that dinosaurs couldn't handle.

This temperature drop was a pretty sudden shift, by geological standards. But it still might have taken many thousands of years to wipe out the dinosaurs, because as Leo points out, you can't measure the effects of climate change from fossils in intervals of much less than 250,000 years.

Not everyone agrees. Jack Wolfe, a paleobotanist with the United States Geological Survey in Denver, Colorado, thinks the weather turned truly awful practically overnight in *T. rex*'s time, so awful that it killed off the dinosaurs.

Jack is one of many scientists who think one or more asteroids struck the earth at the end of dinosaur time, with catastrophic results. To me that's a hot theory in search of evidence. In 1991 Jack added one of the newest wrinkles to the theory, specifying the month when dinosaurs went extinct: June. Jack's findings were published in a science journal and many newspaper stories that summer.

Jack had looked at a lot of plant fossils from the end of dinosaur times in Wyoming. The plants' leaves had opened as if it were early summer. Pollen grains were found in clumps, as they are when plants are not mature. And leaf surfaces were wrinkled, Jack thought, from a flash freeze. But Leo and Kirk noted that the kind of pollen Jack found were dispersed in clumps of four when they matured. And they got a wrinkled texture on leaves by drying them. What's more, they found that the same leaf pattern showed up on leaves from throughout Hell Creek Formation time, which would have to mean multiple flash freezes by Jack's reckoning.

To Jack, the plant wilt was due to a frost—the aftermath of a huge asteroid striking the earth. The impact made a huge crater in the Gulf of Mexico, sending up a cloud of dust that blocked the sun. Just days after the big chill, the air started to warm from carbon dioxide released by the impact with limestone in the sea. The greenhouse effect of the CO_2 increase raised the worldwide temperature ten degrees Fahrenheit, where it stayed for more than half a million years.

At least that's how Jack sees it. Jack's evidence for that change is fossils of leaves that grew three or four times larger than before the end of the dinosaurs, developing sharp tips like tropical leaves today.

To me, all that sounds like a lot to figure out from some fossil plants. And the people who best know the plants of *T. rex*'s time, Kirk Johnson and Leo Hickey, just don't buy it. Kirk and Leo found that what Jack had called frost damage on leaves could just as well have come from rotting and drying out. Kirk and Leo also don't think it's possible to make temperature estimates as precise as

Jack does, and they don't think Jack did enough detailed analysis.

I'm no expert on plants. But I know enough about science to be careful of what I read in the newspapers, and sometimes the scientific journals.

I do know that *T. rex* was around for several million Junes in many environments. *T. rex* roamed only a small part of the dinosaur world, but from what we can tell, *T. rex* encountered several different communities of animals within its range.

CRETACEOUS PARK: *T. REX'S* ANIMAL WORLD

We know a lot more about *T. rex*'s world than we do about most other dinosaur communities. Quite a few of the dinosaurs we know about come from the era of the tyrannosaurs, the last 15 million years of dinosaur life. That's less than 10 percent of dinosaur time. The

richest collections of dinosaurs, however, come from Montana and western Canada at the time of *Albertosaurus*, several million years before *T. rex* lived.

By most estimates (though not all), there were fewer kinds of dinosaurs by the time *T. rex* came along. Over the last several million years of dinosaur time, the weather was changing considerably. The shallow North American interior seaway, the Colorado Sea, appeared about 95 million years ago and receded and advanced several times in the next 30 million years. It was receding right at the end of *T. rex*'s time, when the weather was becoming more continental, drier and cooler in winter, hotter in summer. The uplift of the Rocky Mountains was also creating new weather patterns. More variable weather meant, so it seems, fewer kinds of dinosaurs.

When you sample the dinosaur fossils up through different rock layers from the Late Cretaceous, you can see how the dinosaur community changed. You don't get just different kinds of dinosaurs—that always happens when you move up in time. The dinosaurs you see from *T. rex*'s day were different in other ways. It appears that a lot of the other animals were dropping

out, such as the armored ankylosaurs, the dome-headed pachycephalosaurs, and some other kinds of dinosaurs that were more abundant earlier in the Cretaceous. Either those animals were going extinct or they didn't like the particular environment in the Hell Creek Formation where we find most *T. rex*es.

But *T. rex* wasn't limited to the Hell Creek Formation of Montana and the Dakotas. Just how wide its range was we can't be sure with so few fossils to go by. It is possible that *T. rex* could have ranged far to the north and south of where we have found it so far. We know duckbilled and horned dinosaurs like those we find with *T. rex* in Montana ranged as far north as the Alaskan Arctic in *T. rex*'s time. The Arctic was almost as far north as it is today, and it got dark and cold in winter, even though Arctic weather was more like Seattle's than that of today's Alaskan winters. Browsing dinosaurs might have migrated north each spring for the rich summer plant growth, and *T. rex* might have followed the herds. We don't have any evidence of *T. rex* farther north than Alberta, but we're still looking.

The shrinking, but still substantial seaway across the middle of the continent may have kept *T. rex* from making it to the East Coast. Though we don't have evidence, *T. rex* could have ranged far south of Montana. In the Southwest, *T. rex* would have come across herds of not only horned and duckbilled dinosaurs, but huge four-legged brontosaurs called *Alamosaurus*. Relatives of these giants had ruled North America at the end of the Jurassic period (210 million to 144 million years ago). But in the Cretaceous (144 million to 65 million years ago), they were replaced in North America by the duckbilled dinosaurs. The brontosaurs kept thriving, however, in South America, and by *T. rex*'s day, at least one kind had made its way back into what's now the southwest United States.

Dinosaurs were land creatures that couldn't have swum across an inland sea, but some may have spent a lot of their time around the water. We used to imagine the duckbills as spending much of their time swimming. A lot of fossil evidence showed us that wasn't the case, but we may have gone overboard in taking all duckbills away from water. Some of them did spend a lot of time in watery places, and with their flat feet and flattened tails they could have been good swimmers. We've got footprints that suggest some of their ancestors were poling along river bottoms with their feet.

Paleontologist Don Baird imagines some hadrosaurs were like hippos in their lifestyle. "They'd breed and feed on land but spend lots of time in the water." As Don points out, elephants spend a lot of time in the water. It takes a load off their feet and cools them down—a very economical way to live.

Meat eaters go where the food is, and a water hole, seashore, or riverbank would have been a good place for a scavenging or hunting *T. rex* to hang out. By the sea it could have been rivaled by an even bigger predator, the giant crocodilian *Deinosuchus*. Ned Colbert, a dinosaur paleontologist who used to work at the American Museum of Natural History, has suggested that giant crocodilians could have eaten dinosaurs.

T. rex also lived among, and probably dined on, giant animals that lived inland from the seaway in the western United States. Here the plant eaters included not only duckbills and big horned dinosaurs, but *Leptoceratops,* a primitive-looking horned dinosaur only six feet long and weighing about 120 pounds. Like *Triceratops* and the other much bigger horned dinosaurs, *Leptoceratops* had a parrotlike beak. It had a short frill on the back of its head. But it had no horns, and it might have walked on two legs as likely as on four.

There were other, smaller animals scurrying around *T. rex*'s habitat. Mammals were beginning to increase in size and diversity even before the fall of the dinosaurs gave them room to dominate the world. As the seaways shrank, newly exposed land connections allowed mammals to cross from Asia to North America, as well as from South America to North America, and vice versa. (Late Cretaceous dinosaurs might have done the same, explaining the close resemblance of

T. rex to its Asian cousin, *Tarbosaurus*.)

Though they still grew no bigger than bread boxes, the mammals of *T. rex*'s world occupied many different habitats. In the trees, rodentlike mammals used their choppers to eat the fruits and leaves of flowering trees dinosaurs didn't get to. Marsupials such as the opossum came up from South America and made a good living in the trees and on the ground, eating most anything. The oldest-known primate, a little insect-eater, comes from the Hell Creek Formation. So does the earliest ancestor of the modern cow and other ungulates.

A few paleontologists have suggested that a wave of highly successful mammal immigrants from Asia ate the North American dinosaurs out of business, or that

THE LAST ICE AGE NOTCHED THESE BADLANDS, AROUND THE *T. REX* SITE IN EASTERN MONTANA, REVEALING LAYERS OF SEDIMENT LAID DOWN MORE THAN 65 MILLION YEARS AGO.

© Pat Ortega 1992

animals crossing continents brought infectious diseases that killed the dinosaurs, or that the spread of flowering plants (flowers didn't evolve until the Cretaceous period) created new resources for little creatures. To me these are interesting guesses, but there is absolutely no proof behind any of them.

Up in the trees in *T. rex*'s world were many more birds than we know from earlier dinosaur times. Ancestors of wading ducks with long limbs waddled in the water, and smaller birds patrolled the shoreline. Out on the water, six-foot long birds that could neither walk nor fly dove for fish. Some of the fish they couldn't have swallowed grew to thirteen feet long. And other water creatures, including the crocodilians, grew even larger. Sea turtles reached lengths of twelve feet, and whale-like mosasaurs (actually relatives of monitor lizards) exceeded forty feet long. Like *T. rex*, they might have eaten whatever they wanted. In the seas, that meant diving birds, sharks, and shelled animals.

But flying reptiles, the pterosaurs, still ruled the air. One was the biggest flying animal ever—*Quetzalcoatlus*. Its wingspan topped forty feet, as wide as a fighter plane.

On land, the dinosaurs remained very much in charge. In *T. rex*'s neighborhood lived two-legged browsing dinosaurs no bigger than us, some with thick-skulled heads (probably for butting each other). The speediest dinosaur in *T. rex*'s world was an *Ornithomimus*, an ostrichlike carnivore that might have been the fastest of all dinosaurs. *Troodon,* the smart little hunter with dextrous hands, keen eyesight, and a big brain, was around too. It might have hunted these animals or snatched eggs and the mousy little mammals that skittered through the underbrush.

But small dinosaurs were not very common in *T. rex* country, at least judging from the fossils we can find. We find many more of the huge dinosaurs (for speculations on why they got so big, see Chapter 8). The giants include some of the armored tanks of the dinosaur world, ankylosaurs. These squat plant-eaters grew

MOSASAURS WERE HUGE SEA REPTILES THAT GREW TO *T. REX* PROPORTIONS AND DIED OUT WHEN *T. REX* AND DINOSAURS DID, AT THE END OF THE CRETACEOUS PERIOD.

© Pat Ortega 1992

to thirty-five feet long. A far larger percentage of the dinosaur population of *T. rex*'s world were herds of *Edmontosaurus* and *Anatotitan*, flat-headed duckbills that grew as long as *T. rex*, more than forty feet.

Eating the plant eaters were other huge carnivores, small only by comparison to *T. rex*. *Nanotyrannus*, loosely called the "pygmy" tyrannosaur was "only" fifteen feet long, and a compact model of *Albertosaurus*, which grew to about twenty to thirty feet long, was on the prowl for meat, too.

Look out over what was *T. rex* country. If you see any animals at all, it's cows, and maybe a coyote, deer, antelope, or badger. *T. rex* would have seen more and larger creatures—most likely herds of horned dinosaurs, such as *Triceratops* and *Torosaurus*. Both of these animals were as big as or bigger than any land animal today, more than twenty-five feet long and six tons in weight. They had huge heads with thick frills and horns at least three feet long. *Torosaurus*'s head was over eight feet long, the biggest skull of any animal that ever walked the earth.

Imagine a lush, warm lowland thick with these horned dinosaurs, herding by the thousands under the watchful eyes of a hungry *T. rex*. I wouldn't want to live there then, but it would be a heck of a place to visit.

LEFT: THERE WAS AN ANIMAL BIGGER THAN AND PROBABLY JUST AS SCARY AS *T. REX*. *DEINOSUCHUS*, A CROCODILIAN AS MUCH AS FIFTY FEET LONG. IT MAY HAVE SNAPPED UP DINOSAURS ALONG SHORES.

BELOW: *TROODON*, THE SMARTEST KNOWN DINOSAUR, NO BIGGER THAN A HUMAN.

CHAPTER 8

LIFESTYLES OF THE HUGE AND FAMOUS

ONCE WE'VE FOUND AS much hard data as we can from our *T. rex* skeleton and all the others, and from *T. rex*'s environment, we're back to speculating about *T. rex* in the flesh, this time with some more reasonable inferences. We're left with an awful lot of pure guesswork, however, since we still don't have a *T. rex* nest, eggs, or much of a young *T. rex*.

It's a reasonable guess to say female *T. rexes* laid eggs, since we have eggs from many other dinosaurs, including carnivores like the man-sized *Troodon*. The eggs probably weren't huge, since even the eggs of huge sauropod dinosaurs like the fifty-foot-long titanosaurs aren't any bigger than cannonballs. And *T. rex* probably laid eggs in clusters, whether in two straight lines like the sauropods, spirals like the hypsilophodontids, or concentric circles like *Protoceratops*.

Mom and Dad *T. rex* might have been good parents. The *Maiasaura* duckbills I found in western Montana apparently kept fermenting vegetation on their nests to warm the eggs. And once their young hatched, the parents brought food to their nestlings and chewed it up before feeding them. Their young were born helpless. We know that from the soft and pitted ends of their bones. And from the trampled eggshell we can tell the young were nest-bound for at least one month (judging by comparative growth rates of fast-growing large modern animals) after they were born. So *Maiasaura* had to be a good parent for its young to survive.

On the other hand, *T. rex* parents could have been uncaring. In the same environment where I found

Maiasaura eggs, I also found the bones of the young of a far smaller plant-eater, *Orodromeus makeli*. These little "mountain runners" had smooth, finished bone surfaces even before birth. That means they were probably born up-and-running. Their parents didn't have to give them much care. Maybe baby *T. rex*es were pretty self-sufficient.

But if you asked me to guess, I'd say baby *T. rex*es hung around the nest for a while after they hatched. Maybe the kids even picked up some feeding behaviors tagging along with their parents.

We don't know how long *T. rex* could have lived, nor how big it could have grown, since it continued growing all its life. We can't even say for sure the eleven reasonably well-preserved *T. rex*es we have, all within 10 percent of each other in size, were adults, though chances are they were mature animals. For one, it's difficult to define just what an adult animal is—is it a certain size, or sexually mature? How do we know with a fossil animal?

Perhaps *T. rex* lived more than a hundred years. The

ORODROMEUS WAS A
SMALL PLANT-EATING
DINOSAUR. THE EMBRYOS
AND HATCHLINGS I'VE
FOUND IN MONTANA
SUGGEST THAT IT DIDN'T
REQUIRE THE PARENTING
ATTENTION THE DUCKBILL
MAIASAURA GAVE ITS
YOUNG.

biggest animals tend to be the longest lived today. But few animals make it past a century. From better-known members of *T. rex*'s family we can guess that as a *T. rex* grew, it became sturdier. Its skull became taller and its snout relatively shorter. The eye socket changed from round or oval to keyhole shaped. A big bar grew partway across the middle of the eye socket under the eye, and the horns over the eyebrows and bones behind the eye became tightly sutured, just as ours do when we are children. But in *T. rex* this didn't seem to happen until it was older.

Do these changes mean *T. rex* became less "cute" as it grew up? That's not as dumb a question as it sounds. One of the great animal behavior scientists, Konrad Lorenz, observed that many animals, including ourselves, exhibit parental behavior in response to certain widely shared anatomical and behavioral features of our young. Human babies and many other young animals have round heads, high foreheads, big eyes, small chins, and uncoordinated movements. These are qualities most of us find "cute."

Baby dinosaurs, at least those of *Maiasaura*, had the same cute features, which may have helped them elicit parental care. Baby tyrannosaurs' round eyes were cute, but not their long snouts. Still, young tyrannosaurs might have been different enough from grown-ups that maybe their parents found them cute and worth caring for.

Whether adult *T. rex*es were compatible with one another is another matter. Pete Larson of the Black Hills Institute asserts that "Sue" has broken and healed-over bones from combat with another *T. rex*. Big animals can be pretty nasty to their own kind. Us, for instance. Or crocodiles, which fight each other viciously, even biting a rival's snout clean off, with fatal results. And Bob Bakker suggests *T. rex*es used their skull bumps to butt each other. Of course, Sue could have gotten its wounds after death, and its other injuries from disease, or just from being brittle and clumsy. Or from sex. Perhaps *T. rex*es fought over access to mates.

How did *T. rex* reproduce? That's not a very scientific

BABY MAIASAURS WERE HELPLESS AND COULDN'T LEAVE THE NEST. I DISCOVERED THIS WHEN I EXAMINED NESTLING BONES AND SAW THEY WERE UNFINISHED.

question, but only because we don't know how to answer it. But one scientist, Beverly Halstead of Britain, did spend a lot of time trying to figure out dinosaur sex. Dr. Halstead did some serious research in vertebrate paleontology, but I'm not sure I'd put his theories on dinosaur mating among his most distinguished work. Halstead died in 1991 in a car accident, but he will not be forgotten by those who saw him simulate dinosaur mating on stage with his companion, Helen Haste (they kept their clothes on, to the audience's relief, I think).

Halstead believed that dinosaurs had the same reproductive anatomy as reptiles and birds have today. These animals' sex organs are hidden inside vents, called cloacas, beneath their tails. The dinosaurs mated, Halstead thought, by positioning their tails so their cloacas lined up. The male organ, scarcely a foot long, would engorge with blood and enter the female's cavity and pass sperm into it.

Halstead thought a male *T. rex* would have mounted a female from the rear, with his front limbs on her shoulders and one hind leg across her back while twisting his tail beneath hers to line up their cloacas. All the while, according to Halstead, the male dinosaur would keep one foot on the ground. Otherwise he might have crushed the female. "Their mating had to be done with great delicacy and great decorum," Halstead told *Omni* magazine for an odd article called "*Tyrannosaurus* Sex." Certainly the dinosaurs with plates and spines on their backs, like *Stegosaurus*, would have had to have been very careful.

There is, however, no evidence to support the idea of such delicate sexual acrobatics by any dinosaur. But there is some evidence to suggest we may be able to discriminate one *T. rex* sex from the other.

How Do You Tell the Sex of a *T. Rex*?

When we compare Kathy Wankel's *T. rex* to the ten others we have, we get some sense of the variation that may have existed among individual *T. rex*es. When you look at human skeletons, you can see a great many differences. No two human skeletons, or two dinosaur skeletons, look exactly alike. And we know that human skeletons vary according to race and by sex as well.

As for *T. rex*, eleven skeletons, all of them seemingly adult, still don't make for a big enough sample to say much about the variety you might find in *T. rex*. But it does seem that the creatures we call *T. rex* fall into two different categories. Some of them, including Kathy's *T. rex*, were relatively slim and delicate, at least for a multiton giant. Others were more robustly built. So some researchers have suggested that there were actually two different kinds of tyrannosaurs we've been calling *T. rex*—that we've gotten it wrong ever since

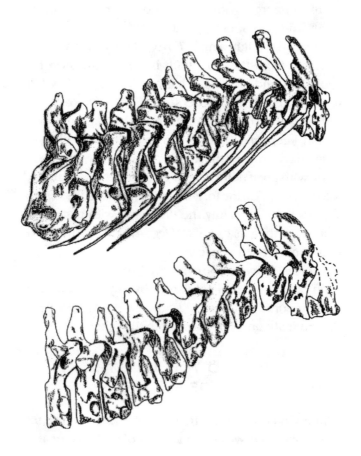

MANY KINDS OF DINOSAURS SHOW SIGNIFICANT DIFFERENCES BETWEEN ADULT SPECIMENS. HERE ARE TWO SETS OF *T. REX* NECK VERTEBRAE, ONE ROBUST, ONE LIGHTLY BUILT. DID ONE BELONG TO A FEMALE? WHICH ONE?

Osborn combined *Dynamosaurus* and *T. rex* into one in 1906.

Dale Russell, a Canadian dinosaur paleontologist, was the first modern scientist to suggest we were calling two different animals *T. rex*. He thought Osborn's two names should have been kept separate. *Dynamosaurus imperiosus* would fit the more massive specimens like the one in Los Angeles; *T. rex* would be just the slighter ones like the skeleton at the Carnegie Museum. Bob Bakker also thinks *T. rex* is two different animals, though he calls them *T. rex* and *Tyrannosaurus "x"*.

Clearly, some *T. rexes* were more robust than others, significantly broader in the head and body. *T. rex* skull expert Ralph Molnar looked at all the *T. rex* skulls except for Kathy Wankel's and Sue. He noticed some differences between them. The one in Los Angeles, for example, had unfused bones in its nose and not much of a bump over its eyes. As I mentioned, on human skulls, and those of many other animals including dinosaurs, skull bones aren't fused in young individuals. Ralph wasn't sure whether the Los Angeles skull was from a different sex or just a younger animal than others he'd examined, even though it didn't vary considerably in size from the other specimens.

The biggest, most noticeable difference between the *T. rex* skulls is the bump over the eyes. In the Los Angeles and Canadian *T. rex* skulls, there is almost no bump at all. In one of the American Museum skulls, the bump is hemisphere shaped. Looking at the bump through a display case, Ralph Molnar thinks it might be a plaster invention. The biggest bump is on the first *T. rex* from the Museum of the Rockies. On that animal the over-the-eye bump is peaked like the eave of a house. So maybe the bump was something that grew on all *T. rexes* as they got older. Then again, it might be a feature of just the females or males. I suspect males had bigger bumps than females, and that the same was the case for features on other dinosaurs that might have been used for display. That's the way it is with horned animals today.

Peter Dodson of the University of Pennsylvania has

done a lot of work on the crests of lambeosaur duckbilled dinosaurs, showing how the crests probably varied within a species according to age and sex. You might figure that the ones with showy crests were the males, who need to display to attract a mate—that's how it is with birds, the descendants of carnivorous dinosaurs, and with rams and deer—but you really can't be sure which dinosaur sex got the headgear.

Ralph didn't find any consistent pattern of variation between the *T. rex*es, judging from just the skulls. But Ken Carpenter did when he looked at *T. rex* skulls and skeletons together. He says the namesake *T. rex* at the Carnegie Museum and Sue are the best examples of the robust *T. rex*. Their jaws are more massive, and they have bigger horns over the eyes and bigger teeth. Kathy Wankel's *T. rex* and those in New York and Los Angeles are the more delicate form.

Ken found several ways in which the *T. rex* skeletons fit into two different groups, from the shape of the arm bones to the number of teeth. For instance, the skinnier, "*gracile*" *T. rex*es like Kathy's and the *T. rex*es in Los Angeles and New York have straight shafts on their upper arm bones. The bigger *T. rex*es like Sue and the namesake *T. rex* at the Carnegie have larger, bowed upper arm bones. (Skeletons of *Tarbosaurus*, *T. rex*'s Mongolian cousin, show the same division between curved-arm and straight-arm-types.)

So maybe there are two types of skeletons of animals we're calling *T. rex*. But I think that before we create two different names for these animals, we have to resolve the question of whether the variation we see is just differences between males and females of one kind of animal, the one we call *T. rex*. That doesn't mean the big ones were males, the thin ones females. That's our prejudice because, on average, men are bigger, women smaller. A lot of mammals turn out that way. But it's not the case for many other animals—many birds, for instance.

As it happens, the female *T. rex* may have been the larger, stronger form. That's what Ken Carpenter thinks. He finds that the bigger *T. rex*es had pubic bones angled

T. REX SKULLS ALSO APPEAR TO SORT INTO TWO CATEGORIES, ONE CONSIDERABLY MORE STOUTLY BUILT THAN THE OTHER. SUE'S SKULL (ABOVE) IS MORE ROBUST THAN THE AMERICAN MUSEUM SPECIMEN (BELOW).

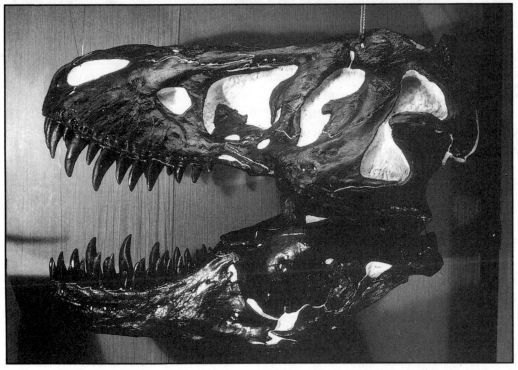

farther away from the back vertebrae. Ken thinks this made for additional abdominal space. That room would have helped eggs to pass through.

Pete Larson of the Black Hills Institute has told me that he and a scientist think they've got a way to figure out the sex of dinosaurs. The scientist is Dr. "Dino," and he's a reptile expert at the natural history museum in Karlsruhe, Germany. Pete compares him to Bob Bakker, a brilliant guy with a wide-ranging curiosity.

From Dr. Frey's nickname you could guess dinosaurs are one of those interests. Frey has dissected a lot of crocodiles in his time. The males are always distinct from the females in the architecture of their tailbones. There is an arch with V-shaped spines on the belly side of the tail vertebrae of reptiles. In anatomy books these are called hemal arches or chevrons. On male crocodilians the first hemal arch, the one closest to the body, is the same size as the second arch. On females the second arch is twice as large as the first. Frey thinks that the male's arch is tight from the muscles that retract the penis and that the female's is enlarged to help it hatch eggs.

Whatever the reason, Frey finds the distinction consistent for crocodiles. And Larson finds it holds up for dinosaurs, too. The first few tail bones of Sue are scrambled up, but those of our *T. rex* at the Museum of the Rockies are in place. The first and second arches are the same size. According to Larson, that would make our *T. rex* a male. And since our *T. rex* is slimmer than Sue, it might be that the more lightly built *T. rex*es were males. Sue, then, would be the right name. (The bones of a subadult found along with Sue would belong to a male.) And Stan, the *T. rex* Pete Larson excavated in 1991, is relatively slim, like our *T. rex*. So it, too, might be male. Our *T. rex*, if it had a name, would have to be named for Tom Wankel, not Kathy.

Yet another clue to the sex of dinosaurs could come from bone studies. Females of many animals, us included, are prone to osteoporosis after menopause. And bone changes during pregnancy can show up in bone

A BIG FEMALE AND A SMALLER MALE *T. REX*? HERE'S ARTIST BRIAN FRANCZAK'S DEPICTION OF TWO SEXES OF *T. REXES*.

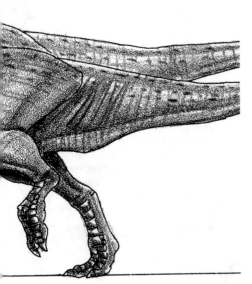

studies. A veterinarian who works in our histology laboratory thinks he may be able to detect these changes in dinosaur bone. If so, then we'd know we had a female on our hands.

But I don't think I'll be convinced there are two species or sexes of *rex*es until we find a lot more than eleven specimens. How many do we need? Maybe thirty. You could take those thirty specimens and compare them to see if they fall into two groups. If they do, and they're all from one vicinity so that the size difference isn't caused by a slightly different gene pool, then the chances are you are looking at two sexes.

Male and female *T. rex*es both were enormous by comparison with most animals, living or dead. Which raises another intriguing question.

WHY DID *T. REX* GET SO BIG?

T. rex wasn't the only giant dinosaur of its day. The herding duckbills and horned dinosaurs were almost as hefty. So I'll rephrase the question: Why were the last dinosaurs so big? Well, maybe it's because the seaway in the middle of North America was drying out and there was more space for bigger animals to browse. And paleontologists have thought for a while that as the herbivores got bigger, the carnivores did, too.

Each group of animals seems to start out small and get bigger over time. Some species, however, remain the same. Those that do eat a lot of different things seem able to adapt to change more readily. Those that grow huge are more limited in the kind and amount of food and the extent of habitat they need, and so are more apt to go extinct than small animals. Throughout the geologic record we see trends toward bigger animals and then the extinction of those big animals. That doesn't mean big animals are more susceptible to extinction. Nearly everything that ever lived is dead. It might have taken a truly catastrophic event to wipe out the last dinosaurs, and the last dinosaurs, big and little, seem to have disappeared at about the same time.

Mammals were no bigger than house cats all through dinosaur times. But when the dinosaurs vanished at the end of the Cretaceous period 65 million years ago, mammals started getting larger until there were giant mammals, bigger than elephants, in some groups in the Eocene (56 million to 34 million years ago). If dinosaurs had lasted longer, a killer even bigger than *T. rex* might have come along.

Canadian paleontologist Phil Currie has a different idea about why *T. rex* and its contemporaries, the last dinosaurs, were giants. The weather was becoming more and more harsh at the end of the Cretaceous. To Phil's thinking, big animals and primitive animals are the best adapted to surviving hard times, at least in the short term. Primitive animals aren't specialized in their behavior and can tolerate a wider range of environments.

T. rex was huge, bigger than almost every preceding carnivore, and far bigger than the meat-eating mammals that replaced it. Some scientists have suggested that tyrannosaurs could have become so big only by

SEA TURTLES ARE THOUGHT OF AS COLD BLOODED, BUT THEIR MASS ENABLES THEM TO RETAIN A CONSISTENTLY HIGH BODY TEMPERATURE. I SUSPECT DINOSAURS BENEFITTED FROM THE HEAT RETAINED BY THEIR MASS AS WELL.

being slow-moving scavengers.

It isn't easy being huge. You've got to keep your population down so you don't use up the available food. But you've also got to keep enough of your kind around to prevent quick extinction. When you apply to dinosaurs estimates based on the number and size of modern mammal predators, as University of Indiana paleontologist Jim Farlow has done, the figures don't work. There wouldn't have been enough *T. rex*es to keep the species alive for the millions of years we know it endured. Something in the world of *T. rex* must have been quite different than it is, say, for lions on the Serengeti Plain today.

Paul Colinvaux, a well-known ecologist, asked and answered the same question in his 1978 book *Why Big Fierce Animals Are Rare.* Colinvaux drew on research that showed, in keeping with the second law of thermodynamics, that energy is lost with each step you take up the food chain. It's a much more efficient use of energy to eat plants than it is to eat the animals that eat plants. The giants of the earth today, blue whales, eat low on the food chain—plankton and shrimplike krill. We don't have many huge animals, Colinvaux figured, because "the energy supply will not stretch to the support of super-dragons."

But *T. rex* was a superdragon. Colinvaux figured it got that way by hoarding its energy:

> The tyrannosaur was not a ferociously active predator... most of its days were spent lying on its belly, a prostration that conserved energy and from which it periodically roused itself.... Nothing like it has been seen since because the true active predators of the age of mammals were able to clean up the meat supplies before a sluggish beast such as a tyrannosaur could get to them. And active predators might even have eaten the tyrannosaur itself.

Jim Farlow accepts Colinvaux's basic premise about energy loss up the food chain limiting the size animals can attain. But he finds the portrait of *T. rex* unflattering and unlikely, and so do I. There were smaller carnivores around in *T. rex*'s day, and we have no evidence they ate *T. rex*. Certainly they didn't wipe it out. And just from appearances, *T. rex* and the other tyrannosaurs weren't built to be sluggish. Ostriches and emus, the large flightless members of *T. rex*'s closest living relatives, the birds, don't do much sitting around.

Think of what factors could have allowed *T. rex* to get huge, instead of what prevents animals from getting big. *T. rex* had some huge plant-eaters to dine on. Who knows what allowed them to get huge? Maybe there was more carbon dioxide in the atmosphere then, as several scientists have suggested. That would have allowed richer plant growth to sustain these gigantic browsers. Maybe duckbills and horned dinosaurs were better at getting energy from their diets than cows, deer, and other modern vegetarians are. A lot of maybes.

You also have to consider metabolism in any speculations of size, diet, and activity levels. If dinosaurian plant-eaters had a slower metabolism than modern mammals, they would have needed to eat less to survive, and so could have lived in greater populations without eating themselves out of existence. If dinosaurs grew slower than mammals do – based on bone studies it seems likely they did in adulthood – dinosaurs probably had a lower metabolism. So, Jim Farlow reckons, dinosaurs could have lived in denser populations than modern mammals can. If dinosaurs' metabolism slowed down when they grew up, there would be even more duckbills to the acre for *T. rex* to eat.

And maybe carnivorous dinosaurs ate less than large mammal predators do. Big animals can maintain a steady body temperature far more easily than little ones, simply by virtue of the size of their bodies. The bigger you get, the more your volume increases in proportion to the surface area of your body. The less surface area in comparison to volume, the less heat flow is pulled away from the body core and lost. So, small animals are more affected by the temperature around them than big animals are. Bigger creatures, even cold-blooded ones

DIMETRODON. NO ANIMAL IS MORE OFTEN WRONGLY LABELED A DINOSAUR. IN TRUTH, IT IS MORE NEARLY RELATED TO US, AND TENS OF MILLIONS OF YEARS OLDER THAN THE OLDEST DINOSAUR.

like sea turtles, can survive prolonged periods of extreme weather because their size and ability to conserve heat, including the heat generated by the movements of their huge muscles, help to keep their temperature stable. This metabolic strategy is called gigantothermy. If *T. rex* were a gigantotherm, it could have kept itself warm on the cold nights, or cool on the terribly hot days that might have killed smaller dinosaurs.

Jim Farlow thinks *T. rex* and other dinosaurs may have had a more variable metabolism. Perhaps they could shift their metabolic rates from low reptilian levels to higher bird and mammal levels according to their age, or the season (a sort of hibernation), or over even shorter time spans. The issue of dinosaur metabolism was a hot one in the 1970s, but was pretty much set aside by the 1980s, with a consensus scientists' answer of "maybe, for some dinosaurs" ("sometimes," Jim Farlow would add).

Now, though, we may be getting some hard answers about dinosaur metabolism. The most interesting new discovery about metabolism doesn't come from paleontologists, but from geochemists—Reese Barrick and Alfred Fischer at the University of Southern California and Bill Showers at North Carolina State. They developed a way of measuring temperature variations within an animal's body, whether it had died recently or hundreds of millions of years ago. When they looked at *T. rex* ribs, backbones, and lower and upper limb bones, the researchers found that little variation had occurred from bone to bone. The results, announced to North American paleontologists at our annual meeting in the fall of 1991, were intriguing to all of us. They suggested that *T. rex* had little internal body temperature variation. Like modern hot-bloods, evidently it had an efficient system for heat retention or maintenance. The same was true of other dinosaurs they sampled. It's not proof that dinosaurs were warm blooded like us. The difference seen in bone temperature might have more to do with how much blood gets out to the limbs than with differing metabolisms. Nor does it mean that *T. rex* was a fast runner, as the researchers have suggested. But it is some of the best hard evidence we have about dinosaur metabolism.

Maybe we can use these measuring techniques to test Jim Farlow's idea that dinosaurs might have had different metabolic strategies at different times in their lives. I've shown with *Maiasaura* and other plant-eating dinosaurs that young dinosaurs grew fast. And the evidence for shifting metabolic rates in dinosaurs is, so far, limited to *T. rex* and hadrosaurs. It does seem that big dinosaurs would have lived very inefficiently if they were hot blooded. Why expend all that energy keeping warm when you can do so by the heat your own bulk produces? Dinosaurs had a lot smaller ratio of body surface area to volume than do smaller animals like

IN *T. REX*'S TIME THERE WERE FISH IN THE SEA, MAMMALS ON LAND, AND BIRDS IN THE AIR, JUST AS TODAY. BUT THE BIGGEST ANIMALS IN THE SEA WERE ICHTHYOSAURS AND PLESIOSAURS, PTEROSAURS IN THE AIR, AND DINOSAURS ON LAND. THIS IS *CTENOTHRISSA*, A LATE CRETACEOUS FISH.

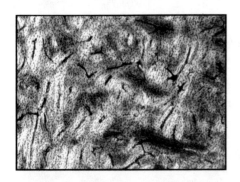

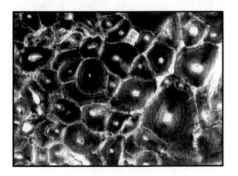

ABOVE: A CROSS-SECTION OF PRIMARY BONE FROM THE LEG OF KATHY WANKEL'S *T. REX*, MAGNIFIED THIRTY TIMES. THE DARK LINES ARE THE VASCULAR OPENINGS FOR BLOOD AND NERVES.

BELOW: AS THE BONE CELLS MATURE, THE PRIMARY BONE IS EATEN AWAY AND REPLACED WITH SECONDARY BONE WITH MORE DENSE AND REGULARLY SHAPED BONE CELLS, ALSO SHOWN AT THIRTY TIMES MAGNIFICATION.

ourselves. That's not to say they couldn't have led high-energy lives as hot-bloods. It just doesn't seem like an efficient strategy for dinosaur life.

There are other prospects for one day figuring out *T. rex*'s metabolism and how it grew so big. If you had a clear, consistent reading of the ratio between the population of predators and prey (or scavengers and potential carrion animals) in *T. rex*'s day, you might be able to make an informed guess about *T. rex*'s metabolism, population, and range. Bob Bakker finds the ratio of dinosaur predators to prey parallels that of modern warm-blooded predators to their prey: about one to three predators for every hundred prey animals. Cold-blooded contemporary predators, like crocodiles, which require far less food, can live in densities ten times as high in relation to their prey. But I don't think Bob's estimates work for dinosaurs. We have too few fossils and too many complicating factors affecting which animals get preserved in different environments to make a reliable estimate of a predator/prey ratio for dinosaurs, or to draw any conclusions about dinosaur metabolism from those ratios.

We have found some evidence ourselves from Kathy's *T. rex* that suggests *T. rex* was warm blooded, but not as we are. Through a generous gift from one of our donors, Anne Merck-Abeles, our museum has the only dinosaur histology lab in the country. That's a facility for studying dinosaur bone microscopically.

We've punched a few little holes in the *T. rex*, in places where it won't make much difference to anyone. From these cores we've cut razor-thin cross-sections. We stain these to show up well under the microscope and then put them in a special viewing microscope and camera at anywhere from $10\times$ to $100\times$ magnification. Then we can take pictures, in black-and-white or color, of what we see.

What we see in *T. rex* (and in many other dinosaurs) are two types of bone development. Both look a little like rounded hollow pasta, especially after we optically color them red in our computer.

The rings, or holes, are sometimes round, but are more often elongated. These are cross-sections of vascular canals for blood vessels. The number of vascular canals suggests dinosaurs had an abundant blood supply, as warm-blooded animals today do. The hole, and the collagen bone tissue that develops around it, are called an osteon. When dinosaurs were young, they had what we call primary osteons. The fibers around the hole were circular and blended in with the surrounding bone. When dinosaurs matured, their bones featured secondary osteons instead. In these the hole expanded, and bone was laid down within the hole.

Dinosaur bone comes in two forms. When young, dinosaurs produced bone in a woven network, much like that of birds and mammals, a pattern we call fibrolamellar. When dinosaurs matured, their bone was laid down in layers, more like the bone of reptiles and amphibians. We call that pattern zonal lamellar. Between each layer of dinosaur bone is a line of arrested growth. Their first arrest line marks the time when I consider dinosaurs adults, though it's not the only possible definition of adulthood, and I don't know how or why that pause in growth happened. Dinosaurs didn't stop growing after that first pause, or any of the many subsequent pauses marked by other arrest lines— dinosaurs grew all their lives. But their rate of growth slowed down considerably after that first pause.

This sort of change happens in our bones about the time we reach forty, though we're long finished growing by then. In duckbills the change in bone structure probably happened when they were just four or four and a half years old. We don't know how old *T. rexes* were when the change in bone structure occurred. We do know that other dinosaurs' growth rates slowed substantially at this time and afterwards, enough for the difference to be recorded in their bones.

So does that mean dinosaurs were hot blooded like us and birds as youngsters, then cold blooded like reptiles as adults? Definitely not. For one, we've been talking about bone-growth patterns, not metabolism. A dinosaur's metabolism didn't shut down when its bones stopped growing, or else it would have died right then. And the bone growth pattern of adult dinosaurs still looks different from that of reptiles. Dinosaurs' bones were fed by more blood vessels than reptiles' bones are.

What the bone studies do show about dinosaur metabolism is that dinosaurs were endothermic homeotherms like us or birds when they were young. After they reached a certain age, or more likely a certain mass, they switched strategies and became mass homeotherms, a bit like the gigantotherms I described earlier. That means dinosaurs' high volume and relatively low surface area, and their ability to utilize external heat and the heat generated by the movement of their muscles, enabled dinosaurs to stay active warm-bloods without burning so much fuel (comparatively) as we and birds must do to stay warm.

These bone studies are the best evidence I know of that dinosaurs had their own special metabolic strategy, unlike anything we know about before or since dinosaurs.

There are other things our bone studies can tell us

THE NARROWNESS OF
T. REX'S SKULL MAY HAVE
ALLOWED ITS EYES TO WORK
STEREOSCOPICALLY, GIVING
IT DEPTH PERCEPTION.

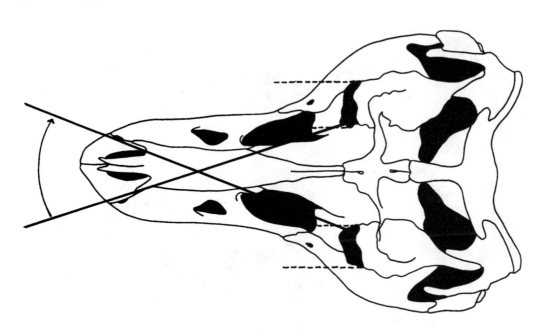

about the life of Kathy's *T. rex* and other *T. rex*es. Sometimes dinosaur bones show changes, like arthritis, that point openly to a well-aged individual. We haven't found that on our *T. rex*. But when we looked at microscopic bone sections of its shin, we saw primary osteons. Why that area had new bone-growth cells we don't know. Maybe the dinosaur had injured its leg and had recovered shortly before it died. But there is no visible sign of an injury on the bone. As you might expect, well-rehealed bone shows up more in young dinosaurs. (If anyone figures that mystery out, it will be my graduate student Mary Schweitzer, who is working on *T. rex* bone histology.)

Perhaps there are other indications of *T. rex*'s metabolism to be found in its skeleton. Bob Bakker, who sees dinosaurs as raging hot-bloods, thinks *T. rex*'s head was ventilated for air conditioning. But a hot head could be a result of *T. rex*'s huge mass, not a hot-blooded metabolism like ours. He also thinks that since some of these holes are located near the brain's blood supply, they might have worked as radiators to cool off *T. rex*'s hot blood, a notion Phil Currie is dubious about. And the

T. REX'S SKULL DESIGN WAS WELL-VENTILATED. WAS THIS FOR AIR-CONDITIONING OR WEIGHT-SAVING?

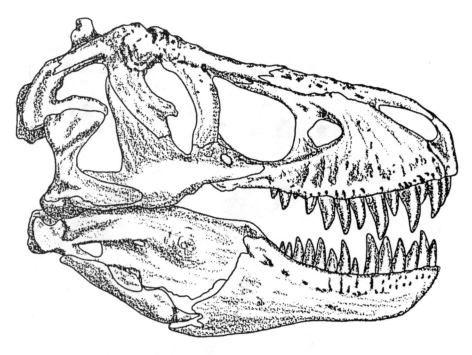

spiral bones discovered in the nose of *T. rex*'s little cousin and contemporary, *Nanotyrannus*, Bob thinks functioned to warm air for a hot-blood. Even if the holes and spirals in *T. rex*'s head weren't radiators, many were conduits for carrying sophisticated sensory information.

How Smart and Sensitive Was *T. Rex*?

T. rex's brain is long gone, but the braincase reveals how large and complicated that brain must have been. The braincase, together with the size and position of the openings for the eyes, ears, and nose, tells us much about how developed the animal's senses were.

When you consider the size of the animal that went with that brain, the brain doesn't seem quite so imposing. Elephants and rhinoceroses have much bigger brains, and these animals weigh less than *T. rex* did. But *T. rex*'s brain was larger than that of almost all reptiles and other dinosaurs; proportionately, it was nearly as big as that of some birds. To be a bird-brain is no compliment by our standards, but for most members of the animal kingdom the comparison is flattering. Some dinosaurs, especially *T. rex*'s small carnivorous cousins, approach the relative brain size of ostriches. Ostriches are far from the smartest birds, but they are among the smartest animals nonetheless. That suggests a lot about the intelligence of those dinosaurs. *T. rex* wasn't as smart as the little carnivores, pound for pound, but it was still bigger brained than the plant eaters of its day.

Relative brain size does seem to correlate to intelligence, though just how smart *T. rex* was we'll never know. Uncertainty hasn't stopped many paleontologists from speculating on its brain power, its nervous system, and what both mean for how *T. rex* sensed its world. It's likely that *T. rex* had superior eyesight for a dinosaur. When you look at the skull, you see right away that its eyes both pointed forward, though not as much as our own. This eye alignment could mean that *T. rex* had depth perception like ours. That's unusual among dinosaurs. If you look at the skull of a *Triceratops*, you'll see its eyes poke out to the side like a cow's. The same with duckbilled dinosaurs.

But you can't tell from a fossil how far the eye stuck out of the orbit. Most birds' eyes are set to the side, but they also stick out far enough to see forward. And just because your eyes point forward doesn't mean they have to work together to give you depth perception. Oilbirds, which do have forward-pointing eyes and overlapping visual fields, don't have stereo vision. *T. rex* also had a deep snout that might have limited the overlap in vision in its eyes. Without other clues to vision that would be found only in soft-tissue parts that don't fossilize, we can't say for sure if *T. rex* could see in three dimensions.

It's even more difficult to say anything about other senses—hearing, taste, smell—until we understand more about the braincase itself. All those holes in *T. rex*'s skull would have been good resonating chambers for transmitting or amplifying sounds. Phil Currie thinks that was their prime function. In birds the holes are associated with the middle ear. Big air chambers behind the eardrum help them hear lower sounds better. *T. rex*

HADROSAURS HAD LARGE NASAL PASSAGES, LIKE THIS *KRITOSAURUS NOTABILIS*.

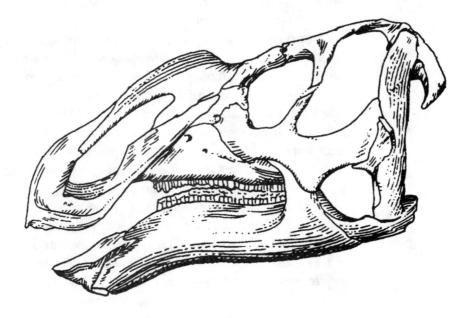

could have used good ears to hear bellowing duckbills from far away.

Bob Bakker has spent a lot of time digging around, literally, in the skull of *T. rex*. He's even taken a coat hanger to probe the pathway of a nerve that runs sideways alongside two other nerves from the braincase to the upper jaw. In humans that nerve gives more sensitivity to touch. In a *T. rex* at the Royal Tyrrell Museum, Bob found that the nerve channel ran straight out of the skull all by itself. Among backboned animals, only birds have the same pattern for that nerve.

Russian paleontologist Leonid Tatarinov suggested the holes in *T. rex*'s snout were nerve openings in what might have been a very sensitive snout. Bob Bakker thinks *T. rex* had a neat row of nerves in the front of the snout that suggest it had lips. Since I know of no counterpart for *T. rex*'s nasal bone structure on any living animal, I won't guess about its function.

Bob with his coat hanger and Phil Currie with needle probes both found that the entire cranial architecture of

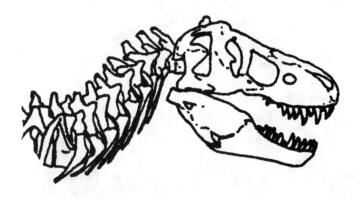

T. REX'S NECK HAD AN EVEN
MORE EXAGGERATED S-CURVE
THAN OTHER DINOSAURS.
PERHAPS THAT GAVE IT MORE
ROOM FOR THE ATTACHMENT
OF MUSCLES TO POWER
RIPPING FLESH.

large carnivorous dinosaurs had a lot in common with small carnivores, and with birds. Bob thinks the oval-shaped holes on each side of the nose of *T. rex* were nerve centers for great sniffing power. A keen sense of smell would be useful for *T. rex*, whether it was a predator or a scavenger, but there's no proving those holes ever had anything in them because cartilage is rarely preserved. But Bob says he's seen the imprint of cartilage on the inside of the snout and compares it to that of modern alligators, which are good smellers. Bob's is a plausible speculation, but those sinuses, and Bob's theory, could just as easily be filled with hot air.

Sometimes Bob isn't going against the facts, just ahead of them. And once in a while the facts catch up and prove Bob's speculation was right after all. That was the case with the nose of that pygmy tyrannosaur, *Nanotyrannus*. The delicate, slightly spiraling turbinal bone within the nostril had never been seen before in a predatory dinosaur. We know turbinals from some keen-smelling mammals like deer. The spiral increases the surface area for better smelling.

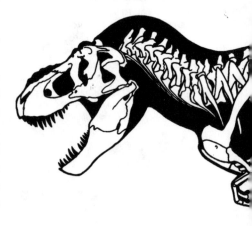

Holes and chambers in the skull can also be useful for resonating sounds. What kind of noises could *T. rex* have produced? Dave Weishampel, a paleontologist at Johns Hopkins University I often work with, has shown how the complicated hollow crests of some duckbilled dinosaurs could have trumpeted low-frequency tunes for great distances. Duckbills were herding animals.

Calling to each other would have been a big help in keeping a herd together and safe from dangers like *T. rex*. It's possible *T. rex* bellowed too, when it had sex or territory on its mind.

Dinosaurs, including *T. rex*, could hear very well, through a special anatomical innovation. But I can't tell you what that is. I'm not being coy. It's just that a graduate student of mine discovered this feature, and until she publishes her research I shouldn't borrow from it.

It's likely that *T. rex* had not only superior senses to most animals of its time, but superior speed. Carnivores usually do, whether scavengers or predators. It does seem *T. rex* was built to move considerably faster than the horned dinosaurs that could have been its prey.

T. REX: HOW FAST, HOW HEAVY?

You can't answer one of these questions without addressing the other. And to make an informed guess

T. REX'S POSTURE WAS LIKE A TEETER-TOTTER, HEAD AND TAIL BALANCED ON EITHER END OF STURDY LEGS.

about either, you need to take a close look at *T. rex*'s complete skeleton, something only lately possible.

When you put *T. rex*'s bones back together, there aren't a lot of ways to get them to fit. You can make a skeleton, and better yet a cast, go wherever you want it to by forcing and bending. A good example is the dramatic new *Barosaurus* mount at the American Museum of Natural History. Defending its young from a hungry *Allosaurus*, that huge plant-eater is rearing up on its hind legs with its neck held high, like Flicka on the old horse movies, only a lot more impressive. It's the tallest dinosaur mount ever. It's also an unlikely pose for a huge dinosaur in life, as the scientists at the American Museum would tell you. And Peter May, who built the exhibit, had to do a fair bit of bending and pulling to get the casts of the bones of *Barosaurus* to fit the desired pose.

As for our *T. rex*, the bones show an S-curve in the neck. Some curve in the neck is a feature common to many dinosaurs—one of the things that sets them apart from other animals. The exaggerated "S" in *T. rex*'s neck drew its huge head farther back toward the hips, a way of pulling the center of gravity back, so the animal wouldn't fall forward.

With the body horizontal, the pelvis had to be aligned parallel to the ground. And the tail, which emerged from the pelvis, would have started out from the body well off the ground. But how did it end up? At our museum, we have only the proximal half of the tailbones from a *T. rex*, the part closest to the body. But we have a clear answer anyway. There were huge transverse processes, grooves and buttresses on the individual tailbones that would have supported large muscles. Those muscles would have been sufficient to keep the tail off the ground. And we see prints of a dragging tail on some of only the earliest of the many dinosaur footprint sites. That's negative evidence, the worst kind, but it still suggests that most dinosaurs kept their tails up.

With its thick tail muscles, *T. rex* would have been able not only to keep its tail well off the ground, but to

swish it around for balance (even use it as a weapon) when it ran. It isn't hard for me to imagine a *T. rex* striding along with its tail held out behind it. You have only to look at crocodiles in the swamp to picture the tail of *T. rex*. Crocs don't drag their tails when they run—they keep them raised. And crocs can go at a pretty good clip, even gallop with a bent-leg waddle. With its long and powerful legs and its balanced two-legged stance, *T. rex* could probably run a lot faster than a crocodile.

With a close look at *T. rex*'s bones and some pretty sophisticated thinking, we can make some reasonable guesses about how fast this dinosaur moved. Estimating how fast *T. rex* might have run is something we'll be able to do better after fully measuring Kathy's *T. rex*. Then we'll be smack in the middle of a good, long-standing argument about dinosaur speed.

When we made *T. rex* the fiercest of a stupid bunch of sluggards, it didn't need to be very fast to survive. Now that so many people think of dinosaurs as hot-blooded speedsters, *T. rex* is the swiftest of the swift—Carl Lewis in lizard skin.

The new image of the hot-running dinosaur was popularized by Bob Bakker. He says that the *T. rex* has

T. REX WEIGHED LITTLE MORE THAN A BULL ELEPHANT, BUT IT WAS CONSIDERABLY LONGER. NEXT TO *T. REX*, THE LARGEST OF THE CATS, THE TIGER, LOOKS LIKE A TABBY.

a large bone surface below its knee, the attachment point for a huge mass of muscle that would have propelled it quickly as a runner. Its thigh bone was indeed massive, seventeen inches around.

Bob then contrasts that thigh dimension with his estimate for the weight of *T. rex*—four tons. That's more bone strength per pound, Bob argues, than a white rhino. Bob says he's driven a jeep in Africa at twenty-eight miles an hour and not put any distance betweeen himself and a white rhino. So *T. rex*, according to Bob, could run forty to forty-two miles per hour. Period.

I think Bob is great for paleontology. He makes a lot of intriguing statements that get other scientists riled up, and sets them to work disproving him. And like much of what Bob has to say, I think his estimate of *T. rex*'s speed is extreme.

Just because something sounds logical doesn't mean it is true. Otherwise, as Ralph Molnar says, we could all be philosophers and not have to grub around in the dirt doing messy research. Scientists look for ways to test a hypothesis like the idea that *T. rex* moved at near highway speeds. Unfortunately, we don't know enough from looking just at bones to be able to tell how fast *T. rex* or any animal moved. It's not good enough to say that because a bone is big, with plenty of room for big muscles, that those muscles and bones were built for speed and not for support or other functions. Nor can you compare a two-legged dinosaur to a four-legged mammal of the same mass and suppose they must have moved at similar speeds. What's more, you can't be sure *T. rex* weighed four tons and not twice as much or half as much. But I'll get to that in a minute.

We can estimate an animal's speed fairly well when we have some other fossil evidence, such as a good set of tracks. If we compare the length of the stride to the size of the animal's leg according to a formula developed by a British zoologist, R. McNeill Alexander, we get a pretty good estimate. Apply his formula to some clear and well-spaced dinosaur tracks in Texas and you

BOB BAKKER CONSIDERS HIMSELF A MAVERICK AMONG PALEONTOLOGISTS, AND LOOKS THE PART.

find that the fastest dinosaur we have footprints for, a medium-sized predator much smaller than *T. rex*, ran twenty-five miles an hour. But we can't say how fast bigger dinosaurs like *T. rex* moved because we have only one footprint from a *T. rex*, identified in northern New Mexico in 1993 by Martin Lockley of the University of Colorado–Denver.

Speed aside, I've never been a great believer in dinosaur footprints as indicators of dinosaur behavior. But many of my colleagues are. They use footprints to draw conclusions about dinosaur herding and hunting as well as speed.

I think we don't know enough about what animal made those footprints, and under what circumstances. Maybe it was one animal using the same path over and over. Who knows if the animal was running at top speed? A *T. rex* probably couldn't sprint over the sort of soft sand or gooey mud that fossil footprints are made from. It may well be that none of the thousands of dinosaur trackways we have were made by a sprinting dinosaur. We don't have any footprints we know for sure came from *T. rex*. So anyone's estimate of how fast *T. rex* moved is just a guess.

Still, trackways do show us things about how dinosaurs moved that skeletons can't. From tracks, we know bipedal dinosaurs walked in an erect fashion, putting

JIM FARLOW STUDIES DINOSAUR FOOTPRINTS, SUCH AS THIS CARNIVOROUS DINOSAUR TRACKWAY HE SUBJECTED TO COMPUTER ANALYSIS.

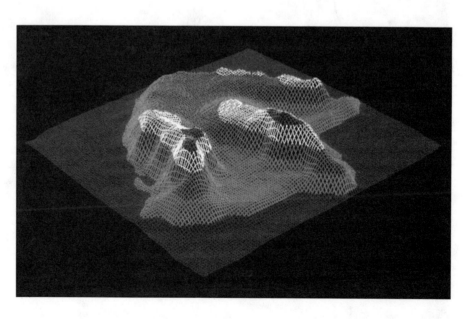

their feet down almost directly one in front of the other.

Jim Farlow, who measured the Texas speedster dinosaurs (probably ornithomimids), does some very careful laboratory studies of dinosaur footprints. He compares them with tracks made by big flightless birds of today, which are very similar to those made by bipedal dinosaurs (of which *T. rex* was one of the biggest). Jim "persuades" emus from the local zoo to run across prepared ground. Then he makes all kinds of measurements of the footprints and converts them to computer graphics. He hopes to use that information one day to tell him how dinosaurs might have distributed their weight while walking and running. And he hopes by studying patterns made by different birds he'll be better able to distinguish among dinosaurs by their footprints. Jim has also tested R. McNeill Alexander's formula for measuring dinosaur speed with his emus. Though Jim's results aren't final yet, videotapes of the emus suggest that they did in fact move at the same speed Jim estimated.

Among living animals the fastest runners are animals in the mid-sized range. Jim's footprint analysis shows the same for dinosaurs, or at least those we have tracks for.

LEFT TO RIGHT: DINOSAUR HINDFEET AND POSSIBLE FOOTPRINT SHAPES: *T. REX*, *IGUANODON*, *APATOSAURUS*.

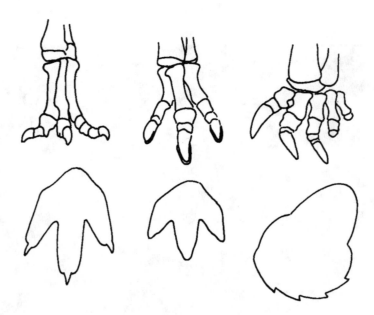

EMUS BEAR MANY
RESEMBLANCES TO MEAT-
EATING DINOSAURS. BUT
COULD *T. REX* HAVE SAT THIS
WAY? I DOUBT IT. I'M
SURPRISED THE EMU COULD.

Mid-sized dinosaurs were the fastest.

But as Jim points out, the fossil footprint record is biased. Many footprint sites are on small patches of land eroded by modern-day rivers. If *T. rex* were moving at twenty-five miles per hour or more, as fast as the fastest mid-sized dinosaurs, its stride would have stretched ten yards or more. There aren't many trackway sites large enough to preserve two such prints. And the squishy mud that takes good footprint impressions isn't a good surface for running fast.

However quickly *T. rex* might have run, it is likely it didn't travel at top speed often. Most animals walk relatively slowly most of the time they are moving, and most dinosaur footprints come from animals just ambling along, not racing.

Without footprints, how can you gauge the speed of *T. rex*? You can get some information from the bones. And if you go just by bones, it is by no means clear that *T. rex* was a fast runner.

Fast-running animals have relatively long legs. They run on their toes, which are few in number and symmetrically arranged (either four or three toes). They have tightly linked foot bones, and their feet and lower legs are long compared with their upper legs. All meat-eating dinosaurs have these features, though *T. rex* doesn't show them as strongly as smaller bipedal dinosaurs do.

But bigger animals may have developed different shapes to support the greater weights of their bodies. If an animal is twice as long as another of the same shape, it will weigh eight times, not twice as much as its smaller version. (Weight is proportional to the cube of the length.) If a *T. rex* needed to be as fast as its mid-sized predator cousin, it might accomplish that, Jim writes, "by making its skeleton stouter, with thicker, more massive leg bones."

T. rex's bones were far stouter than those of the mid-sized predators that were thought to have been the fastest-known dinosaurs. Maybe these heavier limbs are designed to compensate for the stresses associated with fast running. Jim finds that in general, big animals tend to stand, walk, and run in a more stiff-legged, erect way than smaller animals. Jim's compared the angle formed by a leg from where it is when it's farthest back in the stride to where it is when it is swung forward. That angle is far smaller in large animals than it is in small animals. The greater range of motion of the small animals means that they can extend their stride relatively farther when running than a large animal can. To Jim that means *T. rex* probably couldn't extend its stride much. So it probably ran in a relatively stiff-legged fashion.

On R. McNeill Alexander's strength index, *T. rex* ranked along with the elephant. Alexander doesn't think *T. rex* was a fast runner. Not all scientists accept Alexander's formula as reliable. But even those who do subscribe to it have some questions about Alexander's conclusions for dinosaur speeds. Jim Farlow thinks Alexander, as a conservative scientist, may have under-estimated the running abilities of dinosaurs. And the shape of *T. rex*'s body is so different from that of any of the modern animals with which Alexander compared it that using the index to estimate *T. rex*'s speed is a pretty risky business—something Alexander admitted.

The diciest part of estimating *T. rex*'s speed by using Alexander's "athletic index" is getting a good estimate of the mass of the animal. That's easy enough when you have a living creature you can stick on a scale. But how do you weigh a *T. rex*?

You can estimate a dinosaur's mass from the limb proportions themselves. You measure the bone circumference and diameter and multiply that by the kind of bulk modern animals add to their skeletons.

Recently, while a graduate student at Yale University, paleontologist Tom Holtz found that the mass of *T. rex* and other meat-eating dinosaurs can be closely estimated from the application of a mathematical formula to the length of a single bone, the femur. Holtz has found that the tyrannosaurs and the ostrich-mimic dinosaurs (and a few others) are distinct from other meat-eating dinosaurs in the proportions of their legs. Their limbs

KNOWING THE LEG LENGTH
OF A DINOSAUR IS
ESSENTIAL TO CALCULATING
ITS SPEED.

are longer and slimmer near the feet than the legs of other predatory dinosaurs. Because scientists didn't make the scaling adjustment for the slim builds of the tyrannosaurs and ostrich-mimics, Holtz thinks *T. rex*'s mass has been overestimated at seven to eight tons when it should be about four and a half tons. But even Holtz's measurements require an independent estimate of mass. Usually that comes from one source, a toy.

By "toy" I mean an accurate model of *T. rex*. You make a scale model of the bones, then look at the muscle attachments and try and figure out what the musculature of the animal would have been like in life. Maybe you look at large ground-running birds for comparison. Then you put some plastic skin on the model.

When all these features are put together, you can get an estimate of the volume of the model. You do that by measuring the amount of water the model displaces. That's the way Archimedes discovered the concept of measuring volume when he sat in the bathtub, saw the water rise, and yelled "Eureka!"

Since crocodiles, which are fairly closely related to dinosaurs, are about the density of water, Alexander thought it reasonable to assume dinosaurs weighed about what their volume of water weighs. For a model dinosaur, the bath is an immersion in a calibrated container of water. Record the volume of water displaced, convert it to mass, and scale that back up to come up with a measurement of a full-sized *T. rex*'s mass.

The problem with this method is that no one had made an accurate *T. rex* model, in part because before Kathy Wankel found her *T. rex*, no one had anything like a complete skeleton to build from. Now Matt Smith, who sculpted many of the display models in the Museum of the Rockies, has made a model that conforms to the new skeleton (see cover photograph). Jim Farlow, who is now doing the displacement experiment on Matt's model, gets a weight of about six and a half tons from his first of several measurements. I was surprised. This *T. rex* looks so much slimmer than previous ones

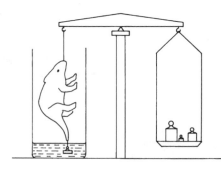

THE BEST ESTIMATES OF DINOSAUR MASS COME FROM SEEMINGLY CRUDE EXPERIMENTS SUCH AS MEASURING THE WATER DISPLACED BY A TOY DINOSAUR MODEL. BUT IF THE MODEL IS GOOD, THE MEASUREMENT CAN BE SOUND.

that I imagined it was going to come out less than four tons. But I accept Jim's conclusions, and his methods, since that's the best we can do for now in estimating dinosaur weight. Since some elephants weigh four tons, and *T. rex* is a far longer animal, Jim's estimate makes sense. And it is still a lot less than the eight tons I've seen listed previously as *T. rex*'s weight.

Our estimates of *T. rex*'s mass are never better than the models we measure from. There's so much uncertainty in this good science that those who study dinosaur speed most closely, like Farlow, say we can't be sure from any anatomical evidence or mathematical projection we have just how fast *T. rex* could run. But if you ask Jim Farlow for his opinion, he'll tell you that he thinks *T. rex* could have run as quickly as twenty-five miles an hour, maybe even thirty-five miles an hour, though Jim doesn't think that likely. Anything faster is highly unlikely. I agree.

For me, it's simply a matter of common sense. To survive, predators need to be only a bit faster or more maneuverable than the fastest animal they hunt. Even if *T. rex* were the most active of hunters, its chief prey would likely have been the most common plant-eaters of its day, *Triceratops* and *Edmontosaurus*. Looking at those bulky animals (in most reconstructions), you can't imagine their moving very quickly. *T. rex* didn't need to be fast to catch a *Triceratops*, so there's no evolutionary reason for an anatomy designed for speed.

Animals try to conserve energy. The bigger an animal is, the more it costs in energy to move its mass. The faster it moves, the more it costs. Furthermore, an animal as big as *T. rex* wouldn't be very maneuverable at high speeds. If it fell, it was a long way down for a head that stood twelve or more feet off the ground. Drop a skull from that height and there's a chance of causing a nasty injury, even a fatal one. And once a *T. rex* fell over, I think it would have had a hard time getting up.

Even at twenty-five miles an hour, *T. rex* is still frightening in my imagination, because at that speed a *T. rex* could still outsprint any of us.

We'll never have that race, but with all we know, and can reasonably guess from our *T. rex* about how that dinosaur sensed, and moved, we can begin to put together a new picture of *T. rex* as a living animal.

T. REX ON THE RUN.

CHAPTER 9

T. REX: PREDATOR
OR SCAVENGER?

WAS *T. REX* A VICIOUS killer? Ask most anyone, including most paleontologists, and they'll say yes. Ask me, I'd say no. You may have noticed that I haven't referred to it as a predator once in this book, only as a carnivore. We're all guessing, but I think those who cast *T. rex* as a predator are letting some common prejudices cloud their thinking.

For sure, *T. rex* ate a lot. It takes a lot of hamburger to feed a six-ton, or bigger, animal. But it shouldn't have mattered to *T. rex* whether lunch was meat caught on the hoof or ripped off a carcass. Either way, *T. rex* had to go get its food. If it were a scavenger, it would have had to roam around looking for carrion. If it were a hunter, it would have had to track down its prey. The chase probably wasn't easy. The big hunters of today, from lions to wolves, miss their quarry nine times out of ten.

What did *T. rex* eat? Pete Larson of the Black Hills Institute thinks he might have the answer in purported *T. rex* fossil scat found at the site of his crew's *T. rex* discovery in South Dakota. Coprolite researcher Karen Chin is going to analyze that material to see what it contains, but she's dubious of its identity. We might be able to determine the diet of the animal that produced it, but there's no way to say for sure that *T. rex* made those coprolites.

A more certain way to determine a dinosaur's diet is to study the isotope concentrations inside the bones themselves. Fossils still have original organic matter in them. If you extract those proteins, you can compare one dinosaur with another. The relative abundance of amino acids in fossils indicates what the animals ate, as geochemist Peggy Ostrom at Michigan State University found, first with sea creatures, and lately with dinosaurs.

Peggy took samples from a lot of dinosaur teeth from the Judith River Formation of Alberta, laid down several million years before *T. rex*'s time. The higher the animal's place on the food chain, the higher the nitrogen content in its bones, or the bones of the animal that ate it. For instance, duckbilled dinosaur bones are not as nitrogen-rich as horned dinosaur bones are. Maybe that's because duckbills ate different plants than the horned dinosaurs did. Peggy isn't sure. She did find that the tyrannosaurids of that time (such as *Albertosaurus*) seemed to prefer to eat duckbills. The carnivores' bones had a level of nitrogen that corresponded to a diet of duckbills, not horned dinosaurs. Smaller dinosaurs, such as the ostrich-mimic ornithomimids, show evidence of a more varied diet.

That doesn't mean *T. rex* liked eating duckbills better. From the Hell Creek Formation it looks like there were a lot of *Triceratops* and duckbills around for *T. rex* to eat, and I don't expect that he was very choosy about what

THIS IS THE LIKELY RESULT OF DINOSAUR'S DIGESTION, A CARNIVORE'S FOSSILIZED WASTE WE CALL A COPROLITE.

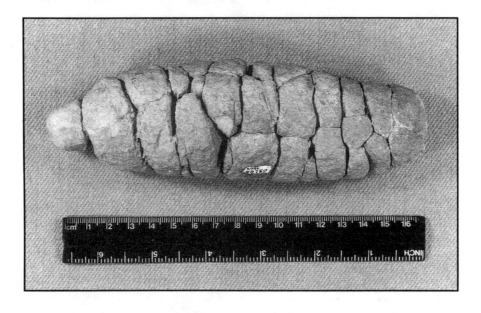

it ate, especially if what it ate was already dead.

How much did *T. rex* eat? Some of the questions of metabolism that enter into that calculation were discussed in the last chapter. Artist Greg Paul has done a lot of speculating about *T. rex*. He figures it would have eaten about four hundred horned dinosaurs if it lived sixty years, or one *Triceratops* every six weeks. Since a *Triceratops* might have weighed twelve thousand pounds, that's more than three hundred pounds of meat a day, maybe twice that. All that *T. rex* feasting, Paul suggests, would have held down the numbers of plant eaters and preserved the vegetation from overgrazing. Paul figures there may have been a quarter of a million *T. rex*es at any one time scrounging around the western floodplain. He estimates that two to five of every hundred dinosaurs that lived then in the West were *T. rex*es.

With an appetite like Paul estimated for *T. rex*, that many *T. rex*es might have eaten one out of every five big plant-eaters every year.

I don't know about that estimate. It assumes *T. rex* was feasting on live young dinosaurs, not on old, sick, and dead ones. And you can't tell populations from percentages of fossils. There are all sorts of built-in biases in the fossil record. Depending on where you look, big dinosaurs may be easier to find than small ones. Certain habitats favor some kinds of dinosaur fossils over others. We get fossils from lowland stream channels, but many kinds of dinosaurs didn't live or die near there.

Accurate diet figures would be based on estimating *T. rex*'s and *Triceratops'* weight, something we can't do very reliably yet, and *T. rex*'s metabolism. We don't know how much running around *T. rex* did or how much energy it burned keeping its temperature steady.

T. rex could have gone most any place in search of a meal. It might well have been a good swimmer. Lots of land animals today are good swimmers, from dogs to horses to ostriches. Greg Paul thinks *T. rex*'s long toes, strong limbs and maneuverable tail would have helped it wade and swim faster than the more hulking prey it might have chased. Canadian paleontologist Phil Currie has found footprints in the Peace River of Alberta that seem to show three mid-sized carnivores chasing some iguanodonts into the water from land. Then again, *T. rex* could have gone into the water just to take a bite of a carcass.

Whether a predator or a scavenger, *T. rex* would have made good use of keen senses to get a meal. Forward-facing eyes, and perhaps stereo vision as a result, would certainly have helped it fix distances when it was tracking down an animal. That visual acuity would help it make precise bites when it was trying to kill, especially in twilight or night, when *T. rex*'s eyesight advantage would be most useful. But acute eyesight would also have helped in spotting distant carcasses. As paleontologist Jim Farlow points out, *T. rex* was capable of standing tall enough to have a big advantage over other dinosaurs in surveying the countryside. A keen sense of smell would help find rotting meat as well as sniff out nearby live prey.

We could just leave it at that—*T. rex* was a hunter or a scavenger, and we'll never know for sure. But it seems to matter a lot to us which *T. rex* was.

Just looking at *T. rex*, most of us would assume it was a hunter, and a darned good one. There are plenty of anatomical indications to suggest *T. rex* was indeed big and powerful: massive legs to chase down prey, teeth half a foot long in huge jaws that could rip off and swallow five hundred pounds of meat in a single bite and gulp. But scavengers can be big and powerful, too.

Paleontologist Don Baird thinks *T. rex*'s size argues for its being a predator, or at least an opportunist that killed when it needed to. Says Don, "There's no evolutionary reason to grow that big if you're a scavenger. Common sense says it evolved that size to be top predator." Top predators today are not the biggest

animals. But some of the biggest animals are scavengers, certainly among the birds, the dinosaurs' closest living relatives.

Historically, *T. rex* was first thought of as the top predator, the king, like the lion today. One of the first news stories ever written about *T. rex*, in the *New York Times* in 1906, refers to a "swift, two-footed 'tyrant,'" so ravenous that its appetite might have explained its very extinction "simply as a result of an insufficiency in the food supply of those days to appease the cravings of his enormous hunger."

Of course we know now that animals don't eat themselves out of business in their own environment. But that journalist's view of *T. rex* was borrowed from the scientist who named it, Henry Fairfield Osborn. "Evidently this was an active, aggressive hunter that relied upon its strong jaws and teeth and upon the heavy claws of its hind legs for bringing down its prey." The man who found *T. rex*, Barnum Brown, said it was "the most formidable fighting machine ever devised by nature."

To folks in the early 1900s, *T. rex* had to be fierce, not just because it looked mean, but because it was old, and therefore primitive and savage. Osborn wrote about "the tendency for the older forms to be the more quarrelsome and wage their combats with greater persistence."

Prejudices about *T. rex* have also run the other way, to thinking of it as a scavenger. In 1917, Canadian paleontologist Lawrence Lambe suggested that *T. rex* and other big dinosaur meat-eaters could only have chewed soft, rotten meat. He based his conclusion on his observation that the crowns of the dinosaurs' big teeth didn't show any wear. We now know that *T. rex*'s teeth were unusually sturdy, and that worn and broken teeth were replaced regularly.

Primitive and savage is how most of us would like to see *T. rex*. That could mean a scavenger as well as a predator. But a predator is more the kind of movie

T. REX AS SCAVENGER AND PREDATOR.

monster that satisfies our need to be scared out of our minds. I think that is why *T. rex* is usually depicted as hunting, running full speed, chasing down an animal, flashing huge teeth, and making bloody kills.

But when you try to figure out how a hunting *T. rex* actually worked, it's still hard to imagine how it caught its dinner. If you look at lions and tigers, animals that catch their prey, you'll see they use their front legs to hold down the victim and their back legs to stabilize it. Then these predators reach over with their mouths and bite and kill their meal. They also don't eat anything nearly as big as they are, like a full-grown *Triceratops* for a *T. rex*.

What *T. rex* did with its arms, nobody knows. In the past, those little arms were commonly assumed to be vestiges. But that doesn't mean they were useless. We have a greatly reduced little toe, but it still is important to us for balance.

Then again, these arms weren't long enough to reach *T. rex*'s mouth, so it can't have used them to put food in its face. Nor could it have used its arms to help it rise from a reclining position. Nearly a century ago, Osborn suggested *T. rex* used its arms to hold onto its date during mating. Jim Farlow thinks *T. rex*'s claws could have assisted its jaws in positioning a mate for sex.

To me, *T. rex*'s using its arms for foreplay beats the heck out of preparator Ken Carpenter's idea. Ken thinks that once *T. rex* caught an animal and brought the victim close to its chest with its jaws, it could use its arms to secure the prey. Ken also examined the hands of Kathy Wankel's *T. rex*, the first ever found intact. He saw that those fingers were positioned very differently from how scientists and artists had guessed in reconstructing *T. rex*. The *T. rex* we see in museums and in books has two hooks set side by side, that would have moved in parallel, like two of our fingers (but not our opposed thumb).

But when Ken and preparator Matt Smith fitted the actual bones together and manipulated them, they found that the claws (if they had been found with the

hand) would have spread apart from the hand. Ken, like Henry Fairfield Osborn, compared the claws to meat hooks with which *T. rex* would have impaled prey and held it fast while lashing out with its arms.

With its big arms, *Allosaurus* could have held onto prey while feeding. Little carnivorous dinosaurs had long and highly mobile arms. But *T. rex* didn't have long arms or a long neck, so it couldn't have held its prey, even a corpse, at a distance with arms outstretched. It needed to get its body up close to its meal for its jaws to do their work. It could have stood on one foot and grabbed its meal with the other foot while tearing off meat with its mouth. I suspect it could have eaten just as well even if its arms were lopped off.

To hold onto its food with its little arms, *T. rex* would have had to have brought its big head way back to take a bite. I don't think that would have been possible, so I don't think Ken's idea is worth a hill of beans.

I certainly don't think those little arms could have been used to catch a living animal, such as a *Triceratops*—by far the most abundant large animal in *T. rex* country. If those hands were useful at all for feeding, and I don't think they were, they'd have been just as good for grabbing onto carcasses the way an eagle uses its legs and claws.

Use your imagination and try to conjure up what it would be like if a *Triceratops* were running down the street and a *Tyrannosaurus rex* tried to catch it. Of course, the *Triceratops* wouldn't want to be caught. If *T. rex* couldn't use its little front legs to catch the *Triceratops*, what could it use?

Well, *T. rex* did have big hind legs. Some have suggested *T. rex* might have run ahead of its prey. Then, with its long tail, *T. rex* could have knocked the animal over. That seems absurd to me. A *Triceratops* would just run off in another direction. Or, perhaps *T. rex* used one leg to jump on its prey. That would mean *T. rex* would have been hopping on the other leg, an action that just

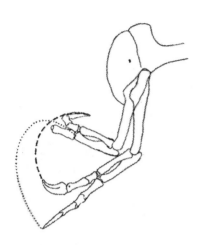

T. REX's ARM MOTIONS WERE SO LIMITED AND ITS ARMS SO SHORT THAT DESPITE THEIR POWER IT IS DIFFICULT TO IMAGINE THEM BEING OF MUCH USE.

doesn't make sense for an animal that large.

Nowadays, you can reconstruct an animal almost any way you want, just as when you go to put plastic flesh on it you can make it fat or skinny. But from what we know about the bone structure of the pelvis of *T. rex*, even if it weighed four tons (at the lower end of scientific estimates), it is very difficult to visualize how it could jump up and down.

We don't have any land animals as big as most dinosaurs today, except for elephants. You can see elephants doing all sorts of weird things with their bodies, including standing on one foot in the circus. But they never jump off their platforms. If an elephant jumped, it would break or dislocate its legs. And until we know more about what happens to dinosaur bone under really stressful conditions, we won't be able to tell whether *T. rex* ever jumped up in the air to get on top of its prey. As for kicking and slashing, the toe claws of tyrannosaurids (as artist Greg Paul has pointed out) were proportionately smaller and duller than those of other meat-eating dinosaurs.

What *T. rex* did have for a weapon was a huge mouth full of enormous teeth. No meat-eating dinosaur had a more powerfully made head than *T. rex*. Those who've studied the skull of *T. rex* see a lot of killing potential to it, but they can't agree on how *T. rex* killed. Greg Paul thinks that *T. rex*'s bite was like a giant "cookie-cutter." Its uniquely thick teeth, arranged in a U-pattern within a powerful jaw, allowed it to snap lumps of flesh a yard wide, down to and sometimes into the bones of its prey. Whereas other tyrannosaurs were slashers, *T. rex* was a chomper, according to Paul. Others who've studied *T. rex*'s jaw come to other conclusions. Paleontologist Bob Bakker says, "The head of *T. rex* was not the head of Godzilla. It was not designed for biting huge chunks of the Chrysler Building." Skull expert Ralph Molnar thinks *T. rex*'s jaws were designed for both slashing and chomping, and opened wide to take especially big bites. To Jim Farlow, a *T. rex* tooth is a fair compromise between a cutting blade, a puncturing tool, and a sturdy support.

It's not easy to find hard evidence for how *T. rex* used its teeth. Broken teeth with smoothed edges have been found in the jaws of several meat-eating dinosaurs. That means the tips of the teeth probably broke off in the process of feeding, and that the teeth were strong enough at the base to stay rooted in the jaw and to get ground down by subsequent chewing.

Paleontologist Tony Fiorillo at the University of California, Berkeley, notes that you don't find tooth marks from any dinosaurs in nearly the same percentage of fossils as you do mammal tooth marks in mammal bone, though tooth-marked dinosaur bone is common. To him, that suggests dinosaurs didn't bite into bone routinely on purpose as hyenas do today.

My former student Greg Erickson has come up with the best evidence yet of *T. rex* teeth marks. He's studied a huge *Triceratops* sacrum found by paleontologist Ken Olson in the Hell Creek Formation in 1990, the first evidence of a *Tyrannosaurus* munching on a *Tricer-*

GREG ERICKSON IS THE FIRST TO STUDY *T. REX* TOOTHMARKS ON BONE, HERE THE SACRUM OF A *TRICERATOPS*. THE *T. REX* LEFT RAKING MARKS THROUGHOUT THE BONE.

atops. Greg also looked at a leg bone of a duckbill and a toe bone of another, both bitten by *T. rex.* How does he know it was *T. rex?* We don't know any other animal from that time and place with teeth big enough to leave such grooves. And the marks are shaped, in relief, just like a *T. rex*'s tooth.

To Greg, the gnaw marks suggest *T. rex* was eating the flesh off bones from meaty places like the butt and calves. Of fifty-eight bites on the sacrum, only two show serrations. To Greg, that indicates *T. rex*'s bite had a forty- to sixty-degree angle of teeth meeting flesh. *T. rex* wasn't sawing the bone, it was puncturing and pulling back.

To me, the bite marks on the sacrum show that this *T. rex* was scavenging an animal it may or may not have killed. It couldn't have gotten its teeth all around a bone four feet long like that sacrum unless the animal was already dead. And if it was trying to kill a *Triceratops,* it would have gone for a more lethal area, like the neck, the muzzle, or the rib cage.

Maybe some other *T. rex* could have reached over

ELEPHANTS CAN DO LOTS OF TRICKS, BUT THEY CAN JUMP ONLY ONCE IN THEIR LIVES. AFTER THAT THEY'D BE CRIPPLED. I SUSPECT *T. REX* WASN'T BUILT TO JUMP EITHER.

and bit into a live adult *Triceratops* as it did in the flip drawings on the facing pages, but I doubt it happened. Even though *T. rex*'s upper teeth were  rounded and seven inches long and do look like they could puncture things, the teeth in the lower jaws were smaller and more knifelike. And all of the teeth, upper and lower, have serrations for cutting flesh. I don't think they could crush the bones of a struggling beast. In this scenario, I can imagine *T. rex* losing some of his teeth in the back of a *Triceratops*. I think *T. rex* was too big and heavy to wrestle around with its prey, delivering bite after bite, the way it did in the movies. If a *T. rex* fell over, it would take a while to get back on its feet, and its prey would be lost.

Maybe *T. rex* shook prey to death. Bob Bakker suggests that the braincase of *T. rex* had lots of room for the attachment of flexing neck muscles. These muscles might have allowed tyrannosaurs to shake their prey violently.

Still, I find it hard to imagine a single bite and a few shakes, even from a *T. rex*, killing or even crippling an animal the size of a *Triceratops*. Maybe *T. rex*'s teeth could puncture some bones, but I don't see how they could puncture a *Triceratops* femur. We do occasionally find grooves on the bones of *Triceratops* that look as if the teeth of a carnivore scratched the bone. I've seen puncture marks on a *Triceratops* femur that were big enough to have been made by a *T. rex*. But they are on both the upper and lower surfaces of the bone. To get its mouth around the big leg bone like that the *T. rex* would have had to have been eating an animal that didn't move—in other words, a corpse. However, Ken Carpenter has a tailbone of a duckbill with puncture marks he thinks come from a *T. rex*. Several spines of the animal's vertebrae were mangled and grew back, so maybe that duckbill survived a *T. rex* attack.

Maybe it lived only to die a more painful death from *T. rex*'s bite days or weeks later, if you buy Bill Abler's

notion of *T. rex*, the infector-killer. Bill Abler is a Chicago researcher on tooth structure in dinosaurs. He thinks *T. rex* infected its victims with bacteria stored in the pocketlike gaps between neighboring serrations in its teeth. Wild as that may sound, it isn't unheard of in the animal world. Komodo dragons, the largest living lizards, grow to nearly ten feet long on the Indonesian island of Komodo. They aren't just big, they're nasty. They will attack animals as big as water buffaloes. Their teeth look somewhat like *T. rex*'s—curved spikes with long rows of serrations on the back edge. But Komodo dragons' teeth are hollow, and their tooth serrations resemble those on leaves, not cubes like *T. rex*'s serrations.

Still, Bill Abler thinks Komodo dragon teeth were enough like *T. rex*'s that *T. rex* may well have killed as Komodo dragons do, by dropping fetid grease and meat particles from their serrated teeth into the tissue of their victims. (Who knows, maybe they had poison glands, too, as do the dinosaurs in Michael Crichton's *Jurassic Park*.)

Before arriving at his novel conclusion, Abler actually went to elaborate lengths to test the cutting action of tyrannosaurid teeth. He built steel blades with different dimensions and serrations. He found *T. rex* teeth had

KOMODO DRAGONS ARE AMONG THE LARGEST COLD-BLOODED KILLERS ON LAND TODAY. THEY KILL BY INFECTIOUS SLOBBER AND HAVE BEEN KNOWN TO CONSUME HUMANS.

more in common with smooth dull blades that act as simple spikes than with hacksawlike serrated blades that grip and rip meat. To grip and rip required coarse serrations like those on the much smaller teeth of the man-sized dinosaur predator, *Troodon*.

As blades, *T. rex* teeth were only marginally efficient. The spaces between neighboring serrations caught and held bits of meat. These chambers could, Abler reckons, have been "havens for the bacteria with which tyrannosaurids infected and subdued their victims." And since *T. rex*'s teeth weren't quite razor-sharp cutters or steak-knife tearers, maybe slow debilitation by poison or blood loss was its best means of weakening a victim. Phil Currie disagrees. "Backed by the strength of its jaws and the weight of the skull, *T. rex*'s teeth are perfect weapons for cutting, processing, and killing."

But maybe *T. rex* did kill, with help. We usually assume *T. rex* was solitary because we haven't found more than one *T. rex* at a time. But we haven't found enough *T. rex*es to say anything knowledgeable about

the animal's social life.

Cooperative hunting, like that in wolf and dog packs today, is associated with pretty high intelligence. But animals a lot less smart than wolves, or *T. rex* for that matter, can hunt together. Some lizards today hunt cooperatively.

Again, we have no hard evidence of cooperative hunting by any dinosaurs. We do have plenty of fossils and trackways to show that plant-eating dinosaurs traveled in huge herds. Bob Bakker believes *T. rex* was a pack animal since some footprints seem to show smaller meat-eating dinosaurs moving in packs. But those marks are from animals a lot smaller than *T. rex*, and could just as easily be left by one or a few animals repeatedly patrolling along the shore.

What makes more sense to me is that perhaps *T. rex* was a scavenger, at least a good percentage of the time. Scavengers aren't stupid, either. Though he saw *T. rex* as a killer, Barnum Brown also wrote that the braincase of *T. rex* "shows a well-developed fore- and hind-brain and abnormally large olfactory lobes. This would indicate that some of the carnivorous dinosaurs at least, depended heavily on their sense of smell when searching for food, and that they were carrion feeders as well as killers."

Life can be simpler and more efficient for scavengers, especially when there are lots of prey animals around. It appears from the number of fossils we've found that *Triceratops* ranged in huge numbers where we find *T. rex*. That means herd animals died from old age, disease, or disaster in big numbers too. If so, *T. rex* didn't have to kill anything on the move. Its meals were already dead.

Or perhaps *T. rex* took down only sickly animals. Scavengers today have a tough time finding enough food in many environments, but in *T. rex*'s day there would have been a good supply of the dead and dying from the huge herds of horned dinosaurs. Even if *T. rex* were a good hunter, why should it have spent expen-

WOLVES HUNT COOPERATIVELY. PERHAPS SOME DINOSAURS WERE SMART ENOUGH TO DO SO AS WELL. BUT I DON'T IMAGINE *T. REXES* DOING SO.

TYRANNOSAURIDS SCAVENGING A MEAL. ALL CARNIVORES, NO MATTER HOW EFFICIENT THEY ARE AS HUNTERS, SCAVENGE AS WELL.

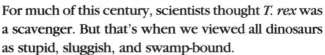

sive energy chasing healthy, live animals when there was plenty of free food to be had?

T. rex as a scavenger isn't a new idea. And it isn't a popular one. For much of this century, scientists thought *T. rex* was a scavenger. But that's when we viewed all dinosaurs as stupid, sluggish, and swamp-bound.

It goes against our prejudices that anything so frightening would not have been a killer. And *T. rex* as a scavenger is unappealing because people think of scavengers, wrongly, as being low on the evolutionary totem pole. We think killers are special and scavengers are slow and stupid. Picking over carcasses is not a behavior we approve of, though we make a party of it every Thanksgiving. The animals that do scavenge, such as vultures, aren't well liked. I think they're graceful myself. But don't think of *T. rex* as a vulture. Think of it as a bald eagle. Bald eagles are mostly scavengers, too.

Scavengers can be fierce as well as clever and alert—they do have to defend their carcasses once they've found them. They can use a big, strong head like *T. rex* had, to process food. And scavengers can better afford the weight of a huge head than can a pure predator, which would need to be more agile to capture its dinner.

I'm not convinced *T. rex* was only a scavenger, though I will say so sometimes just to be contrary and get my colleagues arguing. And I'm not the only scientist who sees *T. rex* as a part-time hunter at most. Another is Jim Farlow. He thinks *T. rex*, like most meateating dinosaurs, was a scavenger that also killed when the opportunity arose. Jim is a cautious scientist, but he's not being just a fence sitter. The way Jim sees it, *T. rex* was tall and had good eyesight, and so could have seen carcasses miles away in forested country. And carrying its head several feet off the ground, *T. rex* could catch an early whiff of the fragrant (to *T. rex* at least) odor of rotting *Triceratops*, since wind speeds increase with altitude, even heights as modest as twelve to fifteen feet. If it came across such a carcass, *T. rex* would have been happy to scarf it up.

On the other hand, attacking a full-grown, long-horned *Triceratops* might have been a good way for *T. rex* to become shish kebab. But a sick, old, or injured *Triceratops* would certainly be fair game. Such weakened animals are the prey among herd animals hunted by many of today's big predators, from lions snaring a baby zebra in the Serengeti to wolves surrounding a sick moose in the Arctic.

Think about the Serengeti for a moment. You've got huge migrations of animals, and thousands of animals dying while crossing rivers, or in floods or droughts. In *T. rex*'s world there were ten thousand, maybe a hundred thousand, horned or duckbilled dinosaurs moving around in herds. Some must have died in just the same way the wildebeest do now. What's going to eat those corpses? If *T. rex* isn't, who the heck is? Talk about a missed opportunity: if *T. rex* wasn't a scavenger then, it would have been very strange. Maybe *T. rex* didn't like rotten meat, but I find that hard to imagine.  I don't know any carnivores that picky. Ever see what a dog will eat? It doesn't matter how many weeks the meat's been lying around, not to my dog, at least.

What about all those scratch marks and broken bones on Sue? What do those tell us? Wasn't *T. rex* fighting prey? *T. rex*es could have busted up each other fighting for carrion, or for mates. Crowded together, battling over a corpse was a good reason for tyrannosaurs whomping each other, stepping on each other's tails and breaking them. A foot bone of the *T. rex* at University of California, Berkeley, is all screwed up as though it got stepped on and rehealed. If "Sue" were a predator, how would it have gone out and gotten food

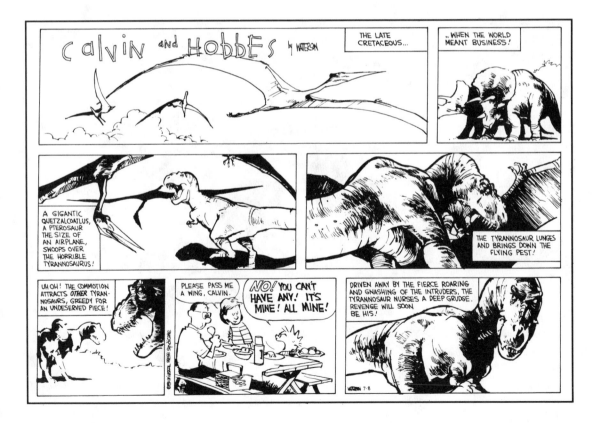

while all those broken bones were healing? A crippled scavenger might have had a better chance of surviving while it was on the mend.

In the end, to me, this whole predator-versus-scavenger debate is a red herring. Most carnivores aren't fussy like us about where they get their meat, whether from dead animals or live ones. Hyenas are scavengers but they don't think twice about killing for lunch, even their own family members. They are opportunists. Among backboned animals, only airborne carrion-searchers like vultures and condors are pure scavengers. The rest take whatever easy pickings they can find, dead or alive. To my mind, *T. rex* was simply the greatest opportunist of them all.

We're lucky to have the opportunity to know *T. rex*, study it, imagine it, and let it scare us. Most of all, we're lucky *T. rex* is dead. And we're not.

VULTURES GET BAD PRESS. THEY ARE ELEGANT, SUCCESSFUL ANIMALS.

MORE
TYRANNOSAURUS REXES?

I think if we spent a couple of years in the badlands of eastern Montana we'd find more *Tyrannosaurus rex* specimens, and other animals we don't even know about yet. But in order to do that you have to have money.

People think we have great piles of money coming in because dinosaurs are so popular, but in fact dinosaur research is one of the least endowed of the biological sciences. Less than a million dollars is spent worldwide on dinosaur research each year. We managed to dig up our *T. rex* on $5,000, but that's because the museum was paying most of our salaries, the Army Corps of Engineers donated equipment, and a lot of people were generous with their time. And it was hard to come up with even that $5,000.

For dinosaur research, most of our funding sources are government agencies, such as the National Science Foundation. And you can't just go to the NSF and say, "Listen, I'd like to go out in the hills of Montana and look for a new dinosaur." You have to have the problem figured out already to get government money.

I spend a lot of my time looking for money so I can do dinosaur research. I do feel if people are making a lot of money from dinosaurs, then at least some of that money ought to go back to research on dinosaurs.

Why study dinosaurs? I do because I love to. I don't expect everyone to share my interest. But there is an awful lot of interest in dinosaurs, especially among children, that justifies working on them. And I think understanding dinosaurs is important. We spend millions finding out what the inside of an atom looks like,

THE FAR SIDE By GARY LARSON

"The picture's pretty bleak, gentlemen. ... The world's climates are changing, the mammals are taking over, and we all have a brain about the size of a walnut."

or trying to see the edge of the universe. And we're out trying to reconstruct the evolutionary biology of the past—what things were really like on this earth. Fossils are the artifacts that tell us about the history of life. They can help us answer questions we want answered, like "How did I get here?" and "How do I fit into the world?"

And we have a hard time getting funding for looking at these questions. I'm not saying it's unfair. But I do think that what we're doing is just as important as seeing the edge of the universe.

I think children deserve the new information we have, information that I don't see on the exhibits in older museums or in the many children's books and toys whose manufacturers don't bother to consult with scientists.

A group of us dinosaur lovers—scientists, artists, and writers—have gotten together to do something about the lack of funds for dinosaur research and the lack of good information in so many dinosaur products. We've formed a nonprofit organization called the Dinosaur Society. The society takes individual and corporate membership funds and spends them on research and education. It publishes a quarterly newsletter and a monthly paper for kids, sponsors trips and digs for the public, recommends scientifically accurate dinosaur books for kids and adults, helps design toys, and advises and endorses worthwhile dinosaur products.

If you'd like to learn more about dinosaurs and what you can do to help dinosaur science, I urge you to write the Dinosaur Society, 200 Carleton Avenue, East Islip, NY 11730

© Pat Ortega 1992

RESOURCE GUIDE

FOR KIDS

Digging Up Tyrannosaurus rex, by John R. Horner and Don Lessem (Crown). How we excavated Kathy Wankel's *T. rex*, with many color photographs.

Tyrannosaurus rex *and Its Kin*, by Helen Roney Sattler (Lothrop, Lee, & Shephard). A brief overview of large carnivorous dinosaurs.

Dino Times. A monthly newspaper for young people, published by the Dinosaur Society, 200 Carleton Avenue, East Islip, NY 11730

FOR KIDS AND ADULTS

The Illustrated Dinosaur Encyclopedia, by Dr. David Norman (Crescent Books). The best review of dinosaur evolution for any age, challenging reading and slightly outdated, but beautifully illustrated by John Sibbick.

The Dinosaur Society Dinosaur Encyclopedia, by Don Lessem and Donald Glut (Random House). The most thorough popular dictionary, with many illustrations.

FOR ADULTS

Digging Dinosaurs by John R. Horner and James Gorman (Workman Publishers). How one paleontologist discovers and interprets dinosaurs.

The Dinosauria, edited by David B. Weishampel, Peter Dodson, and Halska Osmolska (University of California Press). This is an enormous technical book but a very

valuable reference.

Dinosaurs Rediscovered, by Don Lessem (Simon & Schuster). An overview of current dinosaur research worldwide, arranged in an illustrated chronology of dinosaur time.

The Riddle of the Dinosaurs, by John Noble Wilford (Knopf). An excellent history of dinosaur science by a distinguished *New York Times* science writer.

For a complete list of recommended science books, more than sixty in all (out of the hundreds written), chosen by a committee of paleontologists, send $2.50 to the Dinosaur Society, 200 Carleton Avenue, East Islip, NY 11730.

If you'd like to come out to Montana and dig dinosaurs with us, the Museum of the Rockies has programs for adults, families, and for kids only at Egg Mountain in Choteau, Montana. Call 406-994-2251 for information, or write the Museum at Montana State University, Bozeman, MT 59717-0040.

For other dinosaur tours and dig programs around the world, write the Dinosaur Society or call 516-277-7855.

BIBLIOGRAPHY

Wherever possible I've relied on what other scientists tell me when it comes to describing other people's research. I grew up with undiagnosed dyslexia, and reading is not easy for me. Besides, scientific papers are pretty tough going, even for scientists. But I've included some papers that I've read among the publications I've consulted to help me with this book, just in case you're interested in looking through them. It is worth reading at least one scientific paper to see how a scientist organizes and interprets information.

Abler, William L. "The Serrated Teeth of Tyrannosaurid Dinosaurs, and Biting Structures in Other Animals." *Paleobiology*, Volume 18, #2, 1992.

Alexander, R. McNeill. *Dynamics of Dinosaurs and Other Extinct Giants*. New York: Columbia University Press, 1989.

————. "How Dinosaurs Ran." *Scientific American*, April 1991.

Bakker, Robert T. *The Dinosaur Heresies*. New York: Zebra Books,1987.

————. "Inside the Head of a Tiny *T. rex*." *Discover*, March 1992.

————. "The Return of the Dancing Dinosaurs." In *Dinosaurs Past and Present*, vol. 1, edited by S.J. Czerkas and E. C. Olsen. Natural History Museum of Los Angeles County, Los Angeles, CA, 1987.

Bonaparte, J. F., F. E. Novas, and R .A. Coria. "*Carnotaurus sastrei*." *Contributions in Science*. Natural History Museum of Los Angeles County, No. 416 (April 4, 1990).

Brown, Barnum. Collected Papers, 1897-1944, vols. 1 and 2. American Museum of Natural History, New York.

Colbert, Edwin H. *The Great Dinosaur Hunters and Their Discoveries*. New York: Dover Publications, 1984.

Colinvaux, Paul. *Why Big Fierce Animals Are Rare: An Ecologist's Perspective*. Princeton, N.J.: Princeton University Press, 1978.

Currie, Philip, and Kenneth Carpenter. *Dinosaur Systematics*. Cambridge: Cambridge University Press, 1990.

Farlow, James O. "Dynamics of Dinosaurs and Other Extinct Giants: Review." *Paleobiology*, Volume 16, 1990.

————. "Estimates of Dinosaur Speeds from a New Trackway Site in Texas." *Nature*, December 24–31, 1981.

————. "Issues in Evaluating Top Speed and Running Ability of Tyrannosaurus." Unpublished manuscript.

————. "On the Rareness of Big, Fierce Animals: Speculations About the Body Sizes, Population Densities, and Geographic Ranges of Predatory Mammals and Large Carnivorous Dinosaurs." *American Journal of Science*, New Haven, Volume 28, February 19, 1992. 234–41.

Janis, Christine M., and Matthew Carrano. "Scaling of Reproductive Turnover in Archosaurs and Mammals: Why are Large Terrestrial Animals So Rare?" *Ann. Zool. Fennici*. Helsinki: Volume 28, February 19, 1992: 201–16.

Leitch, Andrew. "Leave Them Bones Alone." *Discover*, March 1992.

McGowan, Christopher. *Dinosaurs, Spitfires and Sea Dragons*. Cambridge: Harvard University Press, 1991.

Olson, Everett C. "Sexual Dimorphism in Extinct Amphibians and Reptiles." In *Sexual Dimorphism in Fossil Metazoa and Taxonomic Implications*. International Union of Geological Sciences (Stuttgart) series A, no. 1, 1969.

Osborn, Henry. Collected Papers, 1877–1933. Ameri-

can Museum of Natural History, New York.

————— . "*Tyrannosaurus*, Upper Cretaceous Carnivorous Dinosaur (Second Communication)" *Bulletin of the American Museum of Natural History* 22 (1936): 281–296.

Ostrom, P. H., S. A. Macko, M. H. Engel, J. A. Silfer, and D. Russell. "Geochemical Characterization of High Molecular Weight Material Isolated from Late Cretaceous Fossils." *Organic Geochemistry* 16 (1990): 1139–1144.

Paul, Gregory S. *Predatory Dinosaurs of the World: A Complete Illustrated Guide*. New York: Simon & Schuster, 1988.

Preston, Douglas. *Dinosaurs in the Attic*. St. Martin's Press, 1986.

Sheehan, Peter M., David Fastovsky, Raymond G. Hoffmann, Claudia Berghaus, and Diane L. Gabriel. "Sudden Extinction of the Dinosaurs: Latest Cretaceous Upper Plains, USA." *Science*, November 8, 1991.

"Sioux Me, Sue Me, What Can You Do Me?" *Newsweek*, May 25, 1992.

Tanke, Darren. "K/U Centrosaurine Paleopathologies and Behavioral Implications." *Journal of Vertebrate Paleontology* 9, 1989.

Weishampel, David, Peter Dodson and Halska Osmolszka, ed. *The Dinosauria*. Berkeley and Los Angeles: University of California Press, 1990.

Wolfe, Jack A. "Paleobotanical Evidence for a 'June Impact Winter' at the Cretaceous/Tertiary Boundary." *Nature*, Volume 352, August 1, 1991: 420 - 23.

INDEX

Gurche, John, 93

CREDITS

PREDATOR BEGINS HERE

Page 106 - ©Douglas Henderson
Page 110 - ©Layne Kennedy
Page 111 - ©Greg Paul
Page 112 - Photographs ©Greg Erickson
Page 113 - ©Bill Abler
Page 114-115 - ©Matt Smith
Page 116-119 - ©Kit Mather
Page 118-119 - Photograph by Bruce Selyem,
        ©MOR
Page 120-121 - ©Kit Mather
Page 124-126 - ©Deb Perugi 1993
Page 126 - ©Deb Perugi 1990
Page 127-128 - ©Brian Franczak
Page 129 - Photograph by Don Lessem
Page 130-131 - ©Ken Carpenter
Page 132-134 - ©Brian Franczak
Page 135 - ©Deb Perugi 1993
Page 137 - John Karapelou/©1992 Discover
        Magazine
Page 139 - ©Robert Walters
Page 140-141 - ©Brian Franczak
Page 145-146 - Courtesy of Douglas J.
        Nichols, U.S. Department of the
        Interior, Geological Survey, Denver
Page 148 - Dennis R. Bramen, Royal Tyrrell
        Museum, Alberta, Canada
Page 150-151 - ©Douglas Henderson
Page 153 - Kirk Johnson, DMNH
Page 153 - Courtesy of Marjorie Leggitt,
        Denver Museum of Natural History
Page 154-161 - ©Douglas Henderson
Page 162-164 - ©Pat Ortega
Page 165 - ©Robert Walters 1985
Page 167 - ©Robert Walters
Page 168 - ©Donna Braginetz 1989
Page 169-170 - ©Donna Braginetz
Page 172 - ©Ken Carpenter, DMNH
Page 175
Top: Courtesy of Black Hills Institute, S.D.
Bottom: Courtesy of Donald Baird, AMNH
Page 176-177- ©Brian Franczak
Page 178 - Photograph by Robert Redden,
        ©Animals Animals
Page 180-181 - ©Robert Walters 1989
Page 182 - Pat Ortega ©1992

Page 183 - ©MOR
Page 185 - ©Kit Mather
Page 186 - Courtesy of Ken Carpenter, DMNH
Page 188-189 - ©Robert Walters
Page 190 - Courtesy of Ken Carpenter, DMNH
Page 190-193 - ©Greg Paul
Page 194 - Photograph by Rob Stahmer
Page 195 - Photograph by James Farlow,
        University of Indiana/Perdue
Page 196-199 - R. McNeill Alexander ©1989
        Columbia University Press
Page 197 - Photograph by James Farlow
Page 200-203 - ©Robert Walters
Page 204 - Photograph by Karen Chin,
        Courtesy of Canadian Museum of
        Nature, Ontario
Page 206-207 - Mark Hallett ©1992, All Rights
        Reserved
Page 209 - Ken Carpenter, DMNH
Page 211 - ©Gregory M. Erickson
Page 212-214 - Photographs by Zig
        Leszczynski, ©Animals Animals
Page 215 - ©Bill Abler
Page 216-217 - Top: Photograph Stouffer
        Productions,
        ©Animals Animals
Page 217 - Bottom: ©Brian Franczak
Page 219 - ©1990 Watterson, Dist. by
        Universal Press Syndicate.
        Reprinted with permission, All rights
        reserved
Page 220 - Photograph Carson Baldwin Jr.
        1976 ©Animals Animals
Page 223 - ©1985 Far Works, Inc. Reprinted
        with permission of Universal Press
        Syndicate, All rights reserved
Page 205-238 - Flip illustrations by Jessie J
        Flores/ Dinah-Might Activities, Inc.
Page 240 - ©Pat Ortega

COLOR PAGES

Page 1 - Top: © Douglas Henderson
Bottom: ©David Peters
Page 2 - Bottom: ©C.R. Scotese, Dept. of
        Geology, U.T. Arlington, 1992
Page 2-3 - Illustration by Charles Knight, Field
        Museum of Natural History, Neg #
        CK9T, Chicago
Page 3 - Top: ©Greg Paul 1988
Bottom: ©Mark Hallett 1984, All Rights
        Reserved.
Page 4-5 - ©Douglas Henderson
Page 6 - Top: ©Douglas Henderson
Bottom: ©Greg Paul 1986